An Introduction to Moc

THE ISLE OF MAN RAILWAY

Beyer, Peacock 2-4-0T No. 13 *Kissack* of 1910 arriving at Douglas on the 9.30 ex-Port Erin on September 28, 1953. The train comprises a large 'F' coach F49 and a small 'F' coach with four 'G' vans of various types bringing up the rear. The signals include the large double bracket with six arms protecting the entrance to Douglas station. At the time, the locomotive would be carrying the Indian red livery, while the two coaches would be in IMR 'blood and custard'. ISLE OF MAN RAILWAYS

By DAVID LLOYD-JONES

A 'BRITISH RAILWAY MODELLING' SPECIAL PUBLICATION

Published and printed by: Warners Group Holdings plc, The Maltings, West Street, Bourne, Lincolnshire PE10 9PH.

CONTENTS

PREFACE & ACKNOWLEDGEMENTS:

Dedication -
To SHARON
My long suffering wife
& railway widow.

Over the years, there have been many good general books published on the Island's railways. However, none of them go into the Isle of Man Railway's history and its unique rolling stock in any great depth. All except, that is, for the excellent 'Isle of Man Railway' by Mr James Boyd, published by The Oakwood Press. This book (now in three volumes) is considered to be the Isle of Man Railway 'Bible' and covers all aspects of the IMR history in great detail. However, this 'IMR Bible' is aimed more at the Manx historian and not the modeller.

'An Introduction To Modelling The Isle Of Man Railway' was born out of many years of frustration as a modeller of the system. In the past, I have found like many other IMR modellers, that it is very difficult to find more information in the form of detailed drawings and historical notes on the locomotives and rolling stock of the Isle of Man Railway and the former Manx Northern Railway. This book is the culmination of many years research into this fascinating subject.

When I first started this book last year, I was sitting one evening looking at 60-plus blank pages, wondering how on earth was I going to fill them all? But, once I had taken the plunge and got stuck in, I was quite surprised to find how quickly and easily they all filled up. In fact, I actually ended up with almost twice as much Isle of Man Railway material as I really needed – the next big and more difficult problem was what to leave out!

No book of this type is really a solo project and I would like to take this opportunity to thank the many people who have greatly assisted and encouraged me in getting this my first book off the ground and into print. First, David Brown, Managing Editor of *British Railway Modelling* and the publisher, RFWW Publishing for taking my dream and making it reality.

Mr Robert Smith – Chief Executive, and his officers and staff of the Isle of Man Railways. A special thanks must go to Mr Colin Goldsmith – Rolling Stock Supervisor and his 'Workshop boys' down on the IoM steam railway and of course Mr Alan – "all right mate!" – Corlett – the IMR PR Executive, first for dreaming up the highly successful 1993 'Year of Railways' and the assistance he provided with this book.

The members of my local club – the Isle of Man Model Railway Group; who have provided me with loads of inspiration for this book (most of it un-wittingly) especially the rest of the 'A team' – Frank Warburton, Alex Johnson, John Williamson, Miles Sherman and, of course, Nick Pascoe. All six of us spent a couple of years burning many gallons of midnight oil ressurecting Manx OOn3 from the grave in the form of the club's now famous IMR Castletown layout. Barry C. Lane, Trevor Nall, John Cox, Brian Caton, Les Darbyshire, Denys Brownlee, Jim Lawton and David Haines – the other Manx modellers across the water on the big Island who have all helped by supplying some of their material for inclusion in this book.

No Isle of Man Railway modelling book would be complete without a special mention to my good friend and Manx historian Mr Curtis Devereau – who can be best described as a bit of a Manx character.

Finally, my wife Sharon, who has put up with my hobby for all these years, although, I think she has become a secret IoM railway enthusiast on the quiet and has very kindly proof-read all my notes and supplied countless cups of coffee and the all important chocolate biscuits. In return, I will have to paint the outside of our house once this book is completed!

DAVID LLOYD-JONES
Onchan, Isle of Man
May 1996

INTRODUCTION

The main aim of this book is to encourage, inspire and promote modelling the 3' (914mm) gauge Isle of Man Railway. As the title suggests, this book serves only as an introduction and not a complete "how to" model this unique and very interesting Manx railway system. Hopefully, there is sufficient information contained within these pages to get you started and point you in the right direction, as to where it is possible to find other sources of facts and information on the Isle of Man railway to complete the task.

In order to include as much Isle of Man Railway material as possible into these pages, I have purposely omitted basic model railway construction techniques such as how to build baseboards, track laying, electrics, construction of scenery and buildings and so forth. This information is widely available elsewhere in other railway modelling books, videos and magazines.

Narrow gauge modelling has become more popular in recent years, where the short trains, sharp curves and small stations make ideal model railway subjects. There has also been a noticeable move in recent years towards more accurate narrow gauge modelling and larger narrow gauge prototypes such as the extensive 3' (914mm) gauge railways in Ireland and of course on the Isle of Man.

After spending many years in the railway fashion wilderness, the Isle of Man's magnificent vintage railway systems have been enjoying a revival since the highly successful 1993 'Year of Railways' to commemorate the Centenary of the Manx Electric Railway. The excellent advertising campaign for this event certainly captured the imagination of the railway world. The railway enthusiasts invaded the Island during 1993 literally in their thousands, many for the very first time. One taste of this railway Island full of charm, character and atmosphere and they have been totally hooked.

The success continued into 1994 and 1995; this year was designated the 'International Railway Festival' to commemorate 100 years of the 3' 6" (1008mm) gauge Snaefell Mountain Railway. 1996 is another bumper year for centenaries, with no fewer than three to celebrate – The 3' (914mm) gauge Upper Douglas Cable Tramway, the 4' 8½" (1435mm) gauge Douglas Head Marine Drive Tramway and the well known 2' gauge (609.8mm) Groudle Glen Railway.

In 1998, the Isle of Man Railway reaches its 125th year of operation and 'IMR Steam 125' promises to be a very special and grand event indeed. For the uninitiated, the Isle of Man is a small Island situated in the middle of the Irish Sea, exactly half-way between England and Ireland. The Island itself is only quite small being 33 miles long and 12 miles wide, with a population of about 75,000 people, of which approximately two-thirds live in and around the main holiday resort town and capital Douglas. Tynwald, the Manx Parliament has been governing the Island for well over 1,000 years and is the oldest government in the world. The Isle of Man has its own money, which still includes the paper pound note, and its own Post Office and stamps.

It is also famous for having cats with no tails, Manx kippers and being a tax heaven and, of course, there is the famous TT motorcycle races, where the Island almost sinks under the sheer weight of motorbikes and leather.

The Island also boasts an extensive Victorian transport system, which includes horse trams, electric trams and of course a large 3' (914mm) narrow gauge steam railway network which once linked the whole Island right up to the late 1960s.

I was born in 1960 and I am therefore just old enough to remember the complete Isle of Man Railway system in its full glory. In fact, one of my earliest memories is being taken out by my grandfather for our daily walk around my home town of Ramsey in the north of the Island (although, I think I was just a good excuse for him to escape from the wrath of my grandmother for a couple of hours!). As our house was then less than a couple of hundred yards away from the town's steam railway station, this would be our first port of call. My grandfather was a retired marine engineer with more than a healthy appetite for anything mechanical, especially if it was steam powered.

Our visits to Ramsey station would very often coincide with the arrival of the mid-morning service train from Douglas. At the head was one or, if we were really lucky, two of the classic Manx Beyer, Peacock 2-4-0T locomotives, beautifully turned out in a smart Indian red livery with a highly polished brass dome and copper pipework, and what is more, they all had names! No. 5 Mona, No. 8 Fenella, No. 12 Hutchinson, and my favourite engine No.11 Maitland were all regular visitors at Ramsey station in the early 1960s. Sadly, the Ramsey line closed forever on September 6, 1968. Although I have seen many 'foreign' steam locomotives since then on the big Island across the water, the Manx 2-4-0T Beyer, Peacocks are still my first love. My other grandparents also lived on the Island at Laxey. This grandfather worked on the Manx Electric Railway (MER) for many years, as did my uncle. When I was small, my other grandmother and I would often ride the MER either between Ramsey and Laxey or Laxey and Douglas.

The annual Sunday school picnic and school outing would often be by steam train and my last trip on the Peel line was on a school outing in 1968 to Peel itself. On the return trip I was very disappointed because it was not steam hauled. Instead we were put on to the two ex-County Donagal railcars Nos.19 & 20 for the journey back to Douglas. Looking back, this is the first and only time I have ever been on the railcars while they were in service.

The Isle of Man Railway was taken for granted by the locals – it had always been there and probably always would. However, a couple of months after my school trip to Peel, both the Ramsey and Peel lines closed. The delights and memories of the old Isle of Man Railway have been well and truly implanted in my mind.

Thirty years or so later, I still live on the Island, now just outside the main town Douglas. My first love has always been the unique Manx railways, especially the Isle of Man Steam Railway and now I find myself taking my two young children – the next generation – down to see and ride on the steam trains in Douglas railway station at the weekends. Although, the IMR had been rationalised in order to survive and Douglas station has been reduced to half of its original size, it still retains the atmosphere and charm of a real steam railway.

Only one of the 15 Manx 2-4-0T Beyer, Peacocks has been lost in time, the rest are either stored, on display or part of the current service fleet that just happens to include my favourite engine No. 11 Maitland still in the IMR Indian red livery of 1950-60's, which brings many childhood memories flooding back.

If you have got a taste for having a go at modelling the Isle of Man Railway, this book should set you off on the right track. However, by far the best way to capture the correct atmosphere and flavour of this unique Manx 3' (914mm) gauge line is a personal visit to the Island armed with camera and/or camcorder and notebook. If you have never been before, now is the best time to come and discover the Island's five Victorian railway systems, beautiful scenery and tranquil way of life – and the beer is excellent too! At this point, you're probably thinking "He's got shares in the Isle of Man" – well you would be absolutely right,
I am nothing more than a satisfied Manxman!

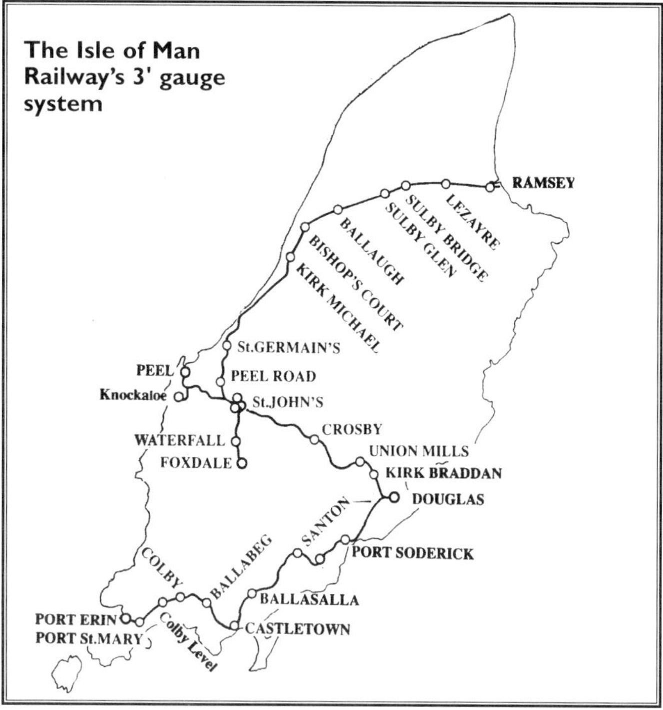

Chapter One

THE ISLE OF MAN RAILWAY IN MINIATURE

The 'Foxdale' train awaits departure on John Cox's freelance Ballabeg station. Gem locomotive No. 10 *G. H. Wood* with F39 'Foxdale" coach. This OOn3 layout was featured in 'Model Railway Constructor'. All buildings are scratch-built and are based on various Manx prototypes.

You may be asking "Why go to the trouble of modelling the Isle of Man Railway, when there are so many other interesting and varied narrow gauge railways already on the UK mainland?" Well, unlike these other narrow gauge railways, the Isle of Man Railway has not been completely over-exposed commercially and offers the narrow gauge modeller new fresh blood, instead of the old favourites such as the Lynton & Barnstaple Railway, Festiniog Railway and the Glyn Valley Tramway.

But, what has the Isle of Man Railway got to offer? Well, this Manx railway is totally unique and is no 'pretend' railway like the many presevered standard and narrow gauge railways dotted up and down the country. The IMR is a real fully working railway absolutely reeking in an atmosphere of smoke, steam and grime. This Manx railway system is unlike any other UK narrow gauge railway as it is principally a passenger railway and was specifically designed to serve the Island's flourishing tourist industry. Huge passenger trains by narrow gauge standards, of up to 14 bogie coaches, were a daily sight. These long trains would be either double-headed or/and banked by the classic icon of the IMR – the elegant Beyer, Peacock 2-4-0Ts.

The whole railway system is locked in a timewarp as a vast marjority of the rolling stock dates back to Victorian times. The last steam locomotive and carriage was delivered to the IMR in 1926, while even today, locomotives from different age are a regular sight on service trains. IMR 2-4-0T No. 4 *Loch* of 1874 is possiably the oldest steam locomotive in world still in regular service, operating on its original line. While, the carriages are virtually un-change from the day they arrived on the Island – electric lighting and seat cushions add that modern 20th century touch!

The genesis of modelling the Isle of Man Railway really started 30 years ago back in early 1960s with several members of the Manchester Model Railway Society. They constructed a huge pioneering IMR layout in 4mm scale running on 12mm gauge track, which became OOn3 or OOn12. This was a very much compressed complete Isle of Man Railway and comprised of all four terminus stations: Douglas, Peel, Port Erin and Ramsey with the junction station of St Johns and both the Foxdale and Knockaloe branches!

The six members lead by Jim Lawton formed the Isle of Man Model Railway Company and constructed the layout privately in six parts, one section in each of the members' homes. The 32' long modular layout was brought together for exhibtions. The whole project was real pioneering stuff on a grand scale. The layout was completed in eight months, which is a remarkable feat, considering at the time there was no commercial support what so ever. Everything had to scratch-built including the engines, while the coaches were made from cut-down Tri-ang clerestories running on TT bogie bolster wagons and the actual IMR wagons were constructed on standard TT chassis.

The numerous appearances on both the exhibition circuit and in the railway press of The Isle of Man Model Railway Company was the first introduction for many modellers to the delights of the Isle of Man Railways. It also marked the start of a period of interest in modelling the Isle of Man Railway with several other Manx layouts appearing on the exhibition scene.

A trilogy of articles on the Isle of Man Model Railway Company featured in the December 1964, July 1965 and March 1966 'Railway Modeller'. A few years later, this historic layout was broken up as space was needed for new projects, although it will be long remembered with great affection by many modellers. It also prompted the first commercially available models of the IMR's locomotives and rolling stock. These were produced in the early 1960s by George E. Mellor (GEM) of 31A Rhos Road, Rhos on Sea, Colwyn Bay, North Wales. The range of GEM Manx kits proved to be very popular with modellers, mainly because the prototype IMR was often featured in the news, as it was in real danger of facing permanent closer in 1965-66 and only managed to avoid this fate at the last-minute.

These white metal kits were to a scale of 4mm to 1', running on 12mm gauge track and consisted of two locomotive kits – the first was the enlarged type Beyer, Peacock 2-4-0T No. 10-No. 13 and first sold for the princely sum of 84/- plus 12/10 for a XT60 motor (a total of about £4.84 in new money). This was followed by the release of the kit of engine No. 16 Mannin with the square-shaped cab. To complete the IOM scene, GEM also produce a small 'F' bogie coach, an 'E' brake van, a 'G' covered van and a 'H' open wagon, also in white-metal. These continued in production right up to the early 1970s, when interest in TT gauge and its narrow gauge derivative OOn3 began to flag with the arrival of the smaller N gauge and OO9. Many narrow gauge modellers were drawn to OO9 and the gimicky 'rabbit warren' or coffee table layouts which could be constructed in a very small area. Instead of getting closer to more accurate scale modelling, narrow gauge took a big step backwards for many years with these glorified train sets. Thankfully, this type of layout has starting to loose its novelity value and now many finescale OO9 and OOn3 layouts are starting to appear.

Although quite crude by modern standards, these GEM OOn3 Manx kits made up into reasonable representations of the prototypes. The main problem was getting the locomotives to run well. The short wheelbase and short circuiting of the pony truck caused many pick-up problems on these little 2-4-0 tank engines.

One simple and effective cure for this problem was to solder the pony truck up solid and file the bottom of the pony truck wheel flat making the engine into an 0-4-0. Another area which cause diffculties was the footplate casting. This was very fragile and prone to break during construction due to the design of the kit.

By the time I advanced from the train-set stage on to kit building with any degree of success, the Gem Isle of Man kits had been long out of production. Scratch-building any Manx rolling stock was for me totally out of the question as I had what seemed like two left-hands and ten thumbs! However, I did manage to find a couple of second-hand GEM items, which included a completed Mannin kit. This had been finished strangely, but very neatly in a British Railways lined black livery complete with the cycling lion emblem! This defaced locomotive was promptly stripped and re-painted into a proper Manx livery. Unfortunately, I could never get it to run well, so it ended up as pride of place on the mantlepiece for many years.

Modelling the Isle of Man and its railways had almost been forgotten along with OOn3 for nearly 25 years, lost in the model railway fashion wilderness. Several years ago, fellow IMR modeller Nick Pascoe and myself tried in vain to revive the interest in Manx modelling by badgering various model manufacturers with scale drawings and photographs in hope they might produce a Manx locomotive and possibly even some rolling stock. Sadly, the usual response from the manufacturers was, "good-looking engines, but there is not enough interest in the Isle of Man and OON3 (4n12) for a viable production run". The only thing that was going to break this stalemate situation was a miniature miracle! Meanwhile, Nick and I spent many summer evenings over two years measuring the remaining IMR carriages in Douglas station as there were not too many good rolling stock drawings, so we could at least get started by scratch-build some rolling stock.

Miracles do occasionally happen – ours came in 1993 in the form of the highly successful "Year of Railways" – by mid-1993, well you know the story – almost every railway enthusiast in the world had heard of the Isle of Man and its wonderful vintage railway systems and about time too! As many railway enthusiasts are also railway modellers, a miniature renaissance of modelling the Isle of Man Railway was bound to follow.

Miraculously, like some fairy tale the spell had been broken – virtually overnight. Nick and I were bombarded with phonecalls, letters and faxes from the very same kit manufacturers wanting much more information on the IMR locomotives and rolling stock, so they could get the kits out on the market as soon as possible to feed the sudden, totally unexpected demand for Manx models.

Roughly about the same time Backwood Miniatures started to produce several first class Irish 3' (914mm) narrow gauge prototypes in 4mm scale – OOn3 was back from the dead after a quarter of century with vengeance!

The first new IMR locomotive kit came out at the end of 1993. This actually consisted of the old 1960s Gem body kits of No. 10-No. 13 and No. 16 with an all new etched nickel silver chassis kit and lost wax brass casting detailing kit from Branchlines of Exeter to bring the old sixties kit up to the much improved standards, both in the level of detail and running expected in the 1990s.

The following summer, after the success of the re-vamped Gem kit – Branchlines released an all new small boilered Beyer, Peacock 2-4-0T kit with enough parts in the kit to construct No. 1 to No. 9 and No. 14 in any condition, apart from No. 1-No. 3 as originally delivered in 1873 with very small side tanks. The kit consists of an all white-metal body with lost wax brass chimney, dome, etc. – while the chassis is etched nickel silver with full brake gear.

Hot on the heels of the locomotive kit, was three etched brass IMR coach kits covering all the large modern (modern on the IMR means up to 1926!) 'F' bogie coaches i.e. F45-F46, F47-F48 and F49. Also, the goods stock has not been neglected either, two all new white-metal kits have been released recently – the two-planked open 'M' wagon and the three-planked open 'H' wagon.

The autumn of 1995 saw Branchlines release the very popular ex-Manx Northern Railway Dübs built 0-6-0 tank Caledonia kit. This kit consists of an all white metal body with lost wax brass chimney, dome, etc. – while the chassis, footplate and buffer beams are etched nickel silver with full brake gear. This locomotive kit has been joined by an IMR cattle wagon and a 'G' covered goods van kit will follow from Branchlines in the very near future.

Meanwhile, another kit manufacturer, Roxey Mouldings, has produced an etched brass kit of almost every IMR and MNR coach type in 4mm scale. Starting with the Foxdale coach F39, closely followed by all the three types of MNR six-wheeled Clemisons and more recently all the different small 'F' bogie coach types covering F1 to F26 and the large saloon coaches F29 to F36, so it is now possible to build a complete rake of both IMR & MNR carriages.

Manx Northern Railway No.1 *Ramsey* crosses a 'landmark' bridge. The bridge, on the Foxdale branch, had an advert for the "Go as you please" tickets painted onto the sides. The lettering could be read very clearly some 30 years after the trains had stopped running to Foxdale. BARRY C. LANE

No. 11 *Maitland* of 1905 leaves Castletown bound for Port Erin, on the Isle of Man Model Railway Group's OOn3 layout. NICK PASCOE

Finally, in 4mm scale Malcolm Mills has produced cast resin kits of the Caledonia and Beyer, Peacock 2-4-0T No. 1 to No. 3 in original condition as supplied to the IMR in 1873.

The level of accuracy and detail on these 3' (914mm) IMR and also Irish 4mm/OOn3 scale kits is amazing, probably making them some of the best narrow gauge kits currently on the market. Remember, until three years ago there was virtually nothing available for any NG modeller wishing to model 3' gauge railways in any scale. There are many new IMR kits in the pipeline, and not all in 4mm scale. 7mm scale is the next target for IMR prototypes.

A live steam Manx small Beyer, Peacock 2-4-0T (No. 1 to No. 9 & No. 14) in 15mm scale, running on Gauge One/45mm gauge track, is available from DJB Engineering. The same company are to produce some carriages and goods stock to go with the locomotive in this scale in near future – a big cheque book is a must for this scale!

If you enjoy and get pleasure out of scratch-building, it is possible to construct the Manx prototypes in several scale/gauge combinations using commercially available ready-built track and associate components such as wheels, motors and gears.

Starting with the smallest possible scale/gauge: 2 mm to 1' (Nn3') – using the ready to run Z gauge Märklin chassis and trackwork it is possible to model a large part or even the whole of the Isle of Man Railway in a moderate sized area. However, if you fancy a real challenge how about scratch-building an entire Manx Beyer, Peacock 2-4-0T locomotive in 2mm scale? Modeller Denys Brownlee has done just that. His 2mm model of the IMR No. 4 Loch in its current condition is a real work of art, accurately capturing the grace and lines of this old Victorian lady perfectly. The model is only a mere 42mm long and is superbly detailed to a level only expected on much larger scale models.

Moving up one scale to 3mm to 1' (TTn3). This scale/gauge utilises the huge range of both British and continental N gauge chassis and ready-built 9mm gauge track. The size of these models would also allow for a more ambitious layout to be constructed in a quite a small area.

However, the most popular scratch-building scale for Manx prototypes seems to be the unlikely scale of 5.5mm to 1', which works out at approximately 7/32". This unusual scale is obtained by dividing the OO/HO 16.5mm track gauge by three. There are a number of good reasons why IMR modellers have chosen to switch to this particular scale/gauge ratio.

Firstly, OO/HO is the biggest model railway market in the world, with vast amounts of readily available wheels, gears, motors and ready-to-run chassis to choose from. Also, many of other OO/HO components such as axlegaurds, roof vents and door handles etc. can be adapted to 5.5 mm scale. Secondly, many modellers have found that due to their small size, the 4mm scale Manx engines did not run reliably enough, especially the early Gem examples. Again, the short wheelbase and short-circuiting of the pony truck caused many pick-up problems on these little 2-4-0 tank engines. In recent years, the new 4mm scale Manx 2-4-0T kits now have modern five-pole motors, flywheels and are controlled by 'smart' feedback controllers, which has dramatically improved their running quality.

The appeal of 5.5mm scale, apart from the huge range of OO/HO items to draw on, is the whole mass and weight of the model is far greater which significantly improves pick-up and the

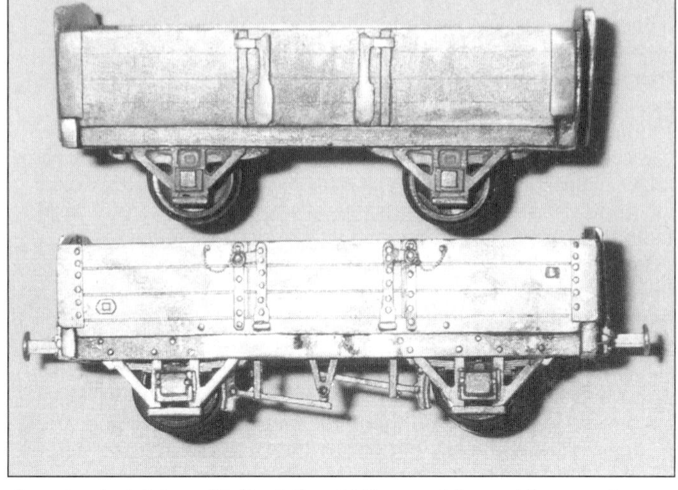

A contrast of styles with (top) the old Gem and (lower) the recent Branchlines 'H' wagon kits, both built and photographed by the author.

power-to-weight ratio – the result is a much smother running model. Also, the 5.5mm models are a nice size and do not require a large area to produce a reasonable size layout.

Two Isle of Man Railway modellers, both from Yorkshire, have gone down the 5.5mm scale road in recent years. After finding OOn3 too small and mechanically un-reliable, Barry C. Lane moved up to 5.5mm scale and has produced a number of very fine Manx layouts in this scale. His 20' long trial Foxdale layout, built in the late 1970s, was exhibited several times between 1978 and 1981.

Foxdale was superceded by the Manx Northern layout in 1980. This also appeared at many events between 1982 and 1985 throughout the north of England. The 18' by 6' layout incorporated many features of the former Manx Northern Railway line from Ramsey to St Johns such as the Glen Wyllin Viaduct, Kirk Michael and Peel Road stations and a tricky section of line called Gob-y-Deigan (Manx for "Mouth of the Devil"). These two layouts, along with Barry's earlier OOn3 Peel station layout, appeared in 'Railway Modeller' August-November 1983.

Brian Caton has a 5.5mm scale layout of St Germain's station on the former MNR line to Ramsey. This layout has been seen at many shows up and down the country and on the Isle of Man in recent years. All the locomotives operating on Brian's layout are live steam. These little 5.5mm steamers are powered by meths and a simple single oscillating cylinder. They all perform faultlessly and the realistic plume of steam out of the chimney adds that finishing touch not possible with electric powered models.

The next scratch-builders scale up the ladder is 10mm to 1' or Gauge One running on O (32mm) gauge track using O gauge wheels, gears and other fittings. A couple of Irish 3' gauge layouts have been built in this scale in the past. This scale/gauge ratio is suitable for use indoors or outside in the garden.

Also very suitable for the garden is 15mm to 1'. Running on 45mm gauge track, this scale has a small amount of trade support from the likes of DJB Engineering, IMP models and similar. This scale/ratio has the potential to take off in a big way! Personally, I like scratch-building in this scale because the parts are nice and big and easy to work on. I have in recent years built a couple of IMR and MNR items in this scale including one of the 'Empress Vans'. Some day, I hope to have a large garden layout based on the IMR with a 12-coach train headed by two gleaming Manx 'Peacocks' running around it.

Any modelling of Manx railways above this scale is really into the realms of model engineering which is normally beyond the skills of the 'average' railway modeller which this book is aimed at.

Crosby station on the Peel line built by Richard Tarpey in OOn3. No.16 *Mannin* makes a rare trip away from the south line with a short train comprising an 'E' luggage/brake van and a pairs coach. RICHARD TARPEY

Douglas to Port Erin trains cross at Ballabeg, John Cox's freelance OOn3 layout, with GEM locomotives No. 4 *Loch* on the left and No. 10 *G. H. Wood* on the right.

The genesis of the IoM model – The Isle of Man Model Railway Company's OOn3 layout. In the foreground, a Peel train is approaching St John's, with *Caledonia* (centre) and a half-brake coach. BRIAN MONAGHAN/RAILWAY MODELLER

Chapter Two

A BRIEF HISTORY OF THE IMR

St John's – the Island's Crewe – in June 1964 with No. 5 *Mona* of 1874 arriving from Douglas while the County Donegal railcars await passengers for Peel. Of interest are the IMR signal and water tank. LES DARBYSHIRE

To model any railway company accurately, first you have to carefully study its history. Most railway companies' histories are long and often very complicated. The rise and fall and revival of the Isle of Man Railway Co. Ltd. is certainly no exception.

To put the IMR in perspective, a brief look at the Island before the setting up and opening of this first Manx railway is necessary.

Up to the early 1800s, the Isle of Man was a sleepy backwater, sparsely populated with no real industries other than local farming, fishing and smuggling – the chief business of the Island!

By the 1850s, things started to improve steadily with the arrival of the Victorian holidaymaker. The Island appealed to the romantic Victorians with its beautiful wild and rugged scenery – plenty of fresh air and long walks. However, outside the four main towns: Douglas, Peel, Ramsey and Castletown (which was still the capital until it lost its title to Douglas in 1869), the rest of the Island was backward and primitive with roads described as dangerous and unworthy of a carriage (some say nothing has changed!).

As the Island's popularity grew each year with Victorian visitors, various attempts were made to build a railway to open the Island's

The last Beyer, Peacock delivered to the IMR in 1926, No. 16 *Mannin*, takes on coal and water at Douglas for the next working to Port Erin. The coal stage at the end of the platforms was used for rapid turnaround. The locomotive was less than a decade old in this 1935 photograph. Mannin was the only engine to carry a square cab from new. The locomotive spent most of its working life on the Port Erin line and is currently on display in Port Erin Museum. MIKE WOOD COLLECTION

IMR No. 1 *Sutherland* of 1873 in rebuilt condition, seen here in Port Erin Museum. One Ross pop valve is situated on the boiler just behind the chimney, the other is in the top of the bell-mouth steam dome. Note the sloping smokebox. The cab spectacle plates are the original round fittings, while the cab front sports rectangular ones on the replacement cab sheet. The snowplough on the floor was fitted during the winter months. It would be nice to see No.1 return to steam for the IMR I25 event in 1998.

interior and link the four main ports. However, it took several attempts over 28 years before the Isle of Man Railway was finally constructed. Starting with the first proposal for an Isle of Man Railway Company was back in September 1845, followed a few weeks later by the Great Manx Railway. A scheme for another Isle of Man Railway in 1857 and again in 1860 were all doomed to fail, as was the Douglas-Peel Railway Company Ltd. in 1862. Another three attempts to promote the Isle of Man Railway Company in 1864, 1867 and 1868 yet again failed to materialise – the main problem with all these schemes seemed to be attracting enough interest to raise the necessary capital.

Finally, a concerted effort was made to open a railway on the Island, with the Isle of Man Railway Co. Ltd. being registered on December 19, 1870 which proposed to connect the Island's three other ports of Port Erin in the south, Peel on the west coast and Ramsey in the north with Douglas. As with earlier schemes money was slow to come in, so the original capital of £200,000 was lowered to £120,000 and the proposed Ramsey line dropped since it was the most expensive to construct and likely to be the least profitable of the three lines.

The appointed Engineer, Henry Vignoles, who had previous experience with narrow gauge railways, recommended a gauge of 3', making the two remaining proposed lines less costly to construct. Messrs. Watson & Smith of London won the contract to build both lines.

The arrival of the railway contractors on the Island in 1872 caused quite a stir. Local labour abandoned its normal traditional employment in favour of a guaranteed three-years work with good rate of pay. In fact, Messers Watson & Smith totally monopolised the Island for several years.

The first line to be completed was Douglas to Peel, which was duly opened on July 1, 1873. Locomotive builders, Beyer, Peacock of Manchester, supplied three 2-4-0 side engines and Metropolitan provided 29 various types of four-wheeled coaches to work the line. The following year, the Douglas to Port Erin line was opened on August 1, for which a further two engines and 27 various four-wheeled carriages were added. The IMR itself had to complete the construction of the south line as Messrs. Watson & Smith were on the point of financial collapse, as the line with its extensive earthworks proved more costly to construct than originally estimated.

The opening of the railway had far reaching social effects on Island life. The Island's interior suddenly found itself in the 19th century virtually overnight, with train loads of Victorian holidaymakers on its doorsteps. The IMR far exceeded the original passenger estimates and surveys and after a couple of years, the IMR could hardly cope with summer traffic that was increasing dramatically each season. A further engine arrived in 1875 from Beyer, Peacock, followed by six large bogie-carriages from Brown Marshalls the following year.

With all the prosperity, growth and wealth the rest of the Island was enjoying, the Ramsey people were determined not to be isolated and formed their own company: the Manx Northern Railway Co. Ltd. in 1877. The most direct route to Douglas from Ramsey was along the Island's rugged east coast. However, due to the terrain, a steam railway was impractical along this route (this shorter route north was exploited by the Manx Electric Railway two decades later). The only alternative was a longer route down the flatter west coast and join up with IMR at St John's station.

The Manx Northern Railway opened a line between Ramsey and St John's on the August 29, 1879. Outwardly, the MNR was very different in many ways to the IMR. Money or lack off it dogged the MNR for its entire 26 year life. Financial constraints often force the MNR to take the lowest tender for equipment. Two 2-4-0Ts were supplied from Sharp, Stewart & Co. of Manchester for the opening of the line. The MNR really needed three engines, but could only afford two. These engines were a plain, workday version of the IMR's Beyer, Peacock 2-4-0Ts. Fourteen six-wheeled Cleminson coaches were also delivered by Swansea Wagon Co. Ltd to work the line. Although the MNR was the IMR's poor relation, the MNR's red sandstone buildings were more substantially built than IMR's original wooden ones. When the Manx Northern opened, clearly a third engine was desperately needed. Sharp, Stewart was unable to supply the engine quickly, so the MNR purchased a Beyer, Peacock 2-4-0T exactly like the IMR ones.

Finally, a line opened between St John's and the lead mines at Foxdale on August 16, 1886 to complete the 3' gauge system. The Foxdale Railway was leased to the MNR. A fourth locomotive, the legendary Caledonia was obtained by the MNR from Dübs & Co. of Glasgow to work the steeply graded Foxdale line. The MNR's involvement in the Foxdale Railway and the Manx Electric Railway, opening its quicker east coast line to Ramsey in 1899, spelt financial disaster for the MNR. Both the Manx Northern Railway and the Foxdale Railway were absorbed by the Isle of Man Railway Co. on April 19, 1905.

The IMR continued to prosper, carrying the ever increasing Victorian and Edwardian holiday makers arriving on the Island. Further enlarged engines and more rolling stock were delivered to combat the heavy loads. To modernise some of the older stock, the original four-wheeled carriage bodies were removed from their underframes and re-mounted in pairs on a new bogie-chassis. This simple conversion gave the IMR 26 'new' bogie carriages (F50 to F75), while the redundant four-wheel chassis were used as a base for a whole host of 'new' goods stock.

The First World War saw the Island swapping roles from a holiday resort to a huge prisoner of war camp. A short special branch was constructed off the Peel line to large camp of 20,000 prisoners at Knockaloe. This temporary steeply graded line was operated exclusively by the ex-MNR engine Caledonia. After the war, Island life returned back to normal with the holidaymakers replacing the internees once again.

It was business as usual for the IMR, although brief competition from the motor bus in the 1920s challenged the IMR's supremacy. The bus war was soon quashed, the railway purchased both its bus

Apart from the modern office block in the background and the square cab of No. 12 *Hutchinson*, this could be an early 1900s shot of Douglas station, before the canopies were added in 1909, a year after No. 12 was delivered. The canopies were removed due to their dangerous state in the late 1970s.

competitors in February 1924, and formed the Isle of Man Road Services. The steam railway and the motor buses now linked every corner of the Island with the main town of Douglas. This early integration of road and rail was to prove a very important factor in the final years of the old Isle of Man Railway Company.

1926 saw the last IMR 2-4-0T No. 16 Mannin delivered from Beyer, Peacock and last coach F49 from Metropolitan. At one stage in the mid-1920s the railway was operating 100 trains daily and carrying a million and half passengers per year. Huge trains by narrow gauge standards of up to 14 bogie coaches would be an everyday sight on the Island during the summer months.

The IMR continued to prosper right up to the out break of war again in 1939. Once again the bucket and spades were replaced by rolls of barbed wire as the Island reverted to being a huge internee camp for the second time this century. Although during the war years, the railway had lost its main source of business – the holiday-maker, these were soon replaced by various service personnel. The War effort kept the IMR very busy indeed, including running more than 14,000 special trains for the military.

After the Second World War, a restoration programme began in 1946 to renovate the IMR ready to tackle the expected traffic as large as pre-war summers. The engines and coaching stock were repainted and new larger engines were consided to replace the tired smaller engines. However, people were slow to return to the Island and the railways saw a steady decline in the vital tourists due to the cheap package holidays to sunny foreign shores, while the arrival of cheaper private motor cars badly hit the local traffic. By the late 1950s, the decline in passenger traffic was serious. Economy drives were begun; late trains and the Sunday service were the first to go. Station were shut, the timetable cut and two diesel railcars were purchased from the former County Donegal Railways in 1961 in an attempt to reduce the ever increasing operating costs.

By 1964, the IMR was in serious trouble. A string of bad summers in the early 1960s had left the IMR with grave financial difficulties. Many other narrow gauge railways around the British Isles had suffered a similar fate, but none had survived as long as the IMR. A touch of 'creative accounting' by the Isle of Man Railway Company headed by the General Manager A. M. Sheard had kept the railway going for so long by seasonally adjusting the figures of the highly profitable Isle of Man Road Services. It was now the buses that was keeping the IMR running.

A. M. Sheard died in July 1965 and the Isle of Man Railway Co. finally succumbed at the end of the 1965 season. No trains ran in 1966, but in 1967 the whole system re-opened, with the Marquis of Ailsa leasing the railway with the assistance from the IOM Government. Unfortunately after two further poor seasons the Ramsey line sadly was closed forever on the September 6, 1967 and the Peel line the following day.

Thankfully the 15½ mile Port Erin line is still operating today. However, its future hung dangerously in the balance in the 1970s, but is now it is solely owned by the IOM Government, along with the Manx Electric and Snaefell Mountain Railways. In the last few years, as a result of the highly successful 1993 Isle of Man 'Year of Railways' the Isle of Man Steam Railway is more popular than ever with both railway enthusiasts and modellers.

Unlike all the other railways around the British Isles, both standard and narrow gauge, the Isle of Man Railway is still the original line, operating the original engines and rolling stock a majority of which dates from the last century. The IMR is still a real railway full of steam, grime and atmosphere.

In 1998 the Island's first railway celebrates 125 years of operation. Another year of special celebrations is planned to mark this milestone in Manx railway history – definitely one not to be missed.

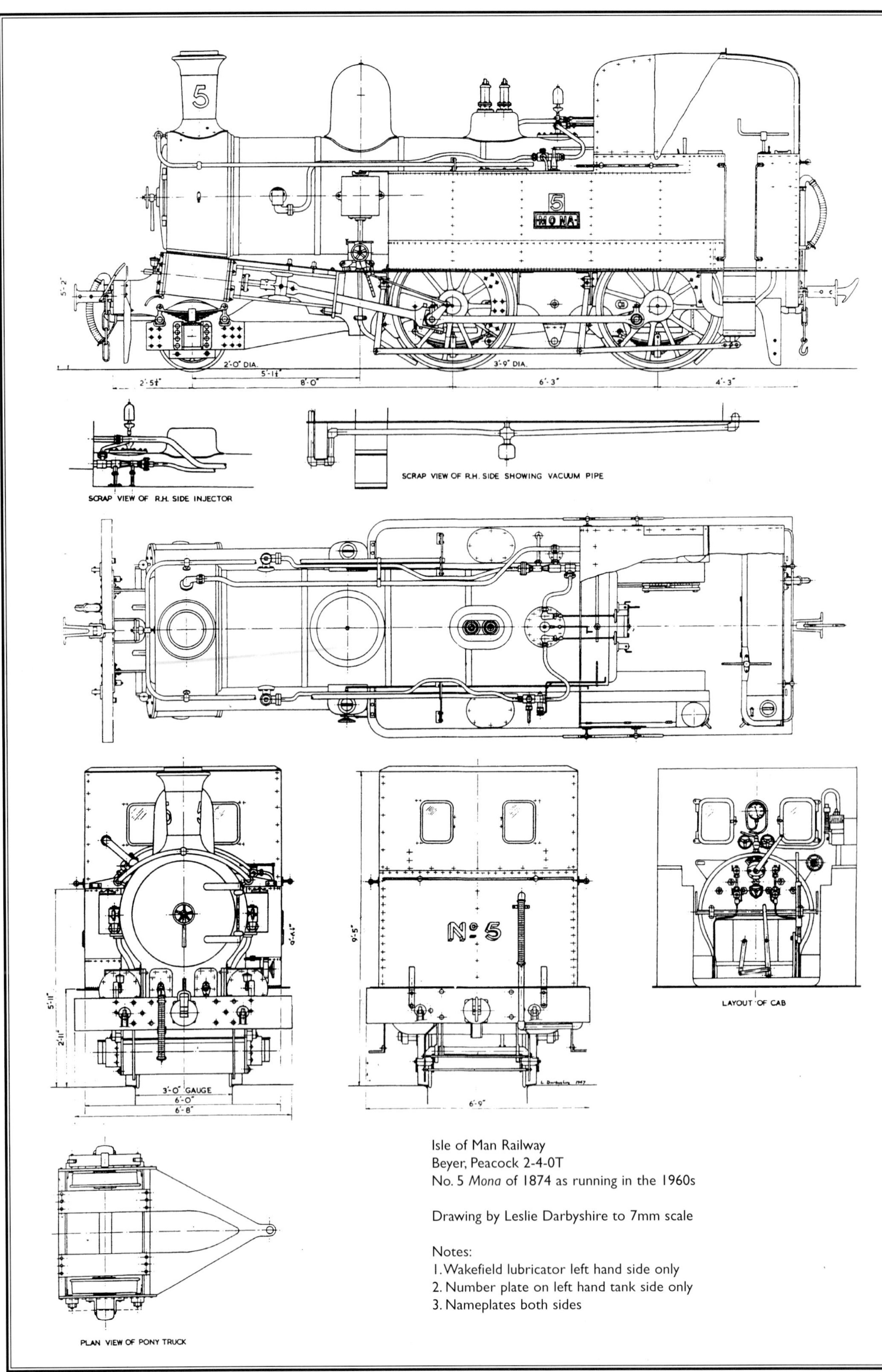

Isle of Man Railway
Beyer, Peacock 2-4-0T
No. 5 *Mona* of 1874 as running in the 1960s

Drawing by Leslie Darbyshire to 7mm scale

Notes:
1. Wakefield lubricator left hand side only
2. Number plate on left hand tank side only
3. Nameplates both sides

Chapter Three

LOCOMOTIVES

The elegant Manx Beyer, Peacock 2-4-0s are well known throughout the railway world and for the enthusiast they symbolise the Isle of Man. A railway's locomotive fleet is normally the main focal point of interest, all other features such as the rolling stock, buildings and signals etc. are usually only of a secondary importance to many enthusiasts. The looks and design of the locomotives are what first attracts many modellers to a particular railway. A desire to discover the rest of a railway's infrastructure usually comes much later, perhaps when thoughts turn to constructing a layout to run the engines on.

A MUSTER OF MANX PEACOCKS

Most modellers like their scale engines to be as accurate and detailed as possible. The history of the Isle of Man steam railway's locomotive fleet is very complicated and needs to be examined in detail to model it with an degree of accuracy. Many railway enthusiasts have said that the Manx engines have a limited appeal as they all look exactly identical. However, a closer inspection will soon reveal they are all in fact very different.

The 15 standard 2-4-0T locomotives supplied to the Isle of Man Railway Co. Ltd. were built by Messrs. Beyer, Peacock of Gorton Works, Manchester over a 53-year period, starting with the first batch of three engines being delivered in time for the opening of the Peel line in 1873.

They are all built to the same basic design, but each subsequent batch being of a larger capacity to counter the increasing traffic demands placed on the railway each year as the Island grew in popularity as the 'playground of the north-west'. This design was not totally unique to the Isle of Man. The origins of these engines can be found in earlier locomotive designs from the same maker such as the Metropolitan Railway's famous 4-4-0Ts of 1864. Messrs. Beyer, Peacock had also previously supplied three very similar looking locomotives to the 3' 6" (1008mm) gauge Norwegian State Railway in 1871. While two engines based on the second batch of Manx 2-4-0s went to the Irish 3' (914mm) gauge Ballymena & Larne Railway in 1877 and 1880 respectfully.

As with any railway company that has been operating the same locomotive fleet for almost 125 years, the Manx 'Peacocks' have obviously been subjected to various alterations during their lifetime. Modifications, enlargements or repairs after accidents have altered the outline of some of these engines. Likewise, boilers, chimneys, domes and many other smaller parts were swapped about regularly in order to keep engines in traffic. The very early standardisation of parts has been a major factor in the railway's survival right up to the present day.

The Manx 'Peacocks' that are in the current operational fleet in the 1990s are a far cry from what was originally delivered. All have had at sometime replacement boilers, side tanks, cabs, wheels and similar. While a couple have even had a new chassis – so all that possibly remains of some of the original locomotives is the name and builder's plates (some of these may well be replicas too!).

As these improvement and repairs took place over many years, it is therefore essential to refer to good photographs of a particular engine within the historical period to be modelled. Followers of the Manx modern image should refer to the monthly railway magazines for reports on the latest developments on the Island's railways. It seems that many new IMR modellers want to reproduce the current scene – reliving memories of a recent visit to the Island perhaps? The last five years has seen some very interesting events take place with IMR engines running on both the Manx Electric Railway and more recently on the Snaefell Mountain Railway. Also, IMR engines that have been in store for many years have been cosmetically restored for display and may one day be overhauled and returned to service once again.

It is now possible to model any of the Manx 'Peacocks' in any variation in 4mm scale due to the recent re-newed interest in modelling the Isle of Man Railway. To a newcomer, the many differences between individual engines can seem very bewildering. Believe it or not, there is a set patten. The main alterations to engines generally occured when the boiler was changed. By

No.12 *Hutchinson* rolls over the Mill Road Crossing, south of Castletown station, with a train for Port Erin on May 3, 1995. The square cab was fitted in 1980.

following the dates on lists of engine boiler changes and the variation notes below and carefullly studying photographs of the locomotives, the pieces of the complex Manx Peacock jig-saw should start to fall into place.

The main variations to be noted are:

Boiler size: various size boilers have been fitted to the engines over the years – varying from the earliest small 2' 10³/₄" diameter boiler – up to the large 3' 7" diameter boiler fitted to two of the engines currently in service.

* See the table on page 27 for a detailed list of individual engine boiler changes.

Brake blocks: All engines were originally fitted with wooden brake blocks, but were replaced with cast iron ones between 1921 and 1924 together with the fitting of steam brakes. The exception was No. 16 of 1926, which was fitted with steam brakes and cast iron blocks from new.

Cab height: Engines Nos. 10-13 were supplied with a 5" higher cab than the original smaller Nos. 1-9 & 14 (ex-MNR loco). Engine No.16 was built with a conventional square-shaped cab and in winter of 1980, engine No. 12 was rebuilt with a square cab similar to that of No. 16.

Cab side canvas: These were fitted to all engines soon after they

The flight deck of 1874 locomotive No.4 *Loch* showing vacuum and steam pressure gauges and vacuum and steam brakes (top right).

arrived on the island and are a very distinctive feature of the Manx engines. Since the early 1970s, all the service engines have had the canvas draft dodgers rolled up on the top of the cab roof.

Cab spectacle plates: Nos. 1-3 were originally fitted with round spectacle plates – new cab fronts with rectangular spectacle plates were fitted to these engines when they were rebuilt with larger boilers and side tanks. No. 2 had a replacement rear cab with rectangular spectacle plates after an accident in 1909. All the later engines were supplied with rectangular spectacle plates from new.

Chimney: All engines were built with a classic Beyer, Peacock tapered chimney with a copper cap. In later years, some engines had cast replacement chimneys – many engines swapped chimneys, so check photographs of individual engines.

Chimney numerals: Each engine had polished brass numerals on either side of the chimney, although as chimneys were swapped around the numerals were removed. Check photographs of individual engines within the given historical period being modelled to see if that particular engine had numerals fitted or not.

Displacement lubricators: Each engine had two brass displacement lubricators either side of the smokebox. In later years these were replaced on certain locomotives with a Wakefield mechanical Lubricator – see notes below.

Handrails: The handrails on the boilers differ from batch to batch, as does the handrail on the back of the cab – there are at least three different variations. Check photographs of individual engines within the given historical period being modelled for the correct type of handrails fitted.

Injector overflow pipe: On engines Nos. 1-3, the original injector overflow pipe end was on the tank tops. Nos. 4-9 were supplied with these pipes extended down the outside of the tanks and through the very narrow running plate at the base of the tanks. Nos. 1-3 were also soon altered to this arrangement. On Nos. 10-13 & 16, the overflow pipes passed through the tanks and out of the base.

Lamp irons: There seems to be some considerable variation in the position of the lamp irons on the leading wooden bufferbeams. Early photographs show them on the front of the bufferbeam and later ones show them positioned on the rear of the beam on some engines – check photographs of individual engines.

Name & builder plates: The locomotives have always carried their nameplates on the tank sides (see main table for details). The only two odd ones are those on No. 4 & No. 5 – they are spelt "LO CH" and "MO NA", the space in-between the first two letters and the last two is to allow the injector overflow pipe to bisect the nameplates. The nameplates on Nos. 1-3 were moved forward when the injector overflow pipes were extended on these engines. Locomotives Nos. 6-9 were supplied with the nameplates and builders' plates either side of the overflow pipes. As engines Nos. 10-13 & 16 had no external overflow pipes, the nameplates are fitted centrally on the tank sides. The builders' plates on these later locomotives are of a smaller type and are fitted on the cab sides.

Oil cans: Just behind the leading bufferbeam, on either side of the coupling housing is a small shelf, which was used to carry an assortment of oilcans and some very old teapots for engine lubrication.

Pipework: There is a slight variation in the shape of the polished copper pipes which connect the injectors on the tank tops with the clack valves on either side of the boiler – check photographs of individual engines.

Safety valves: All locomotives, except No. 16 were originally supplied with Salter safety valves mounted on a tall, highly polished brass bell-mouthed dome. Subsequent re-boilerings of the locomotives over the years has seen some of the engines retaining their Salter valves, while others have been fitted with Ross pop safety valves. Engines Nos. 9 & 14 (ex-MNR locomotive) still retain the original Salter valves. Engines Nos.1-3 & 7 have a single Ross pop valve mounted half-way between the dome and chimney, with a second Ross pop valve mounted in the top of the bell-mouthed dome. Engines Nos. 4-6, 8, 10-13 were rebuilt with a highly polished brass closed dome and two Ross pop valves mounted half-way between the dome and steam turret. Locomotive No.16 was also supplied in this condition. – check photographs of individual engines within the given historical period being modelled for the correct type of safety valves fitted to a particular engine.

A detail shot of No.10 *G. H. Wood* in current condition showing twin Ross pop valves, steam turret with cheeky whistle and injector on tank top with copper pipe leading to clack valves, and black handrail below. The side tank fillers are inside the cab.

A similar view of No.10 *G. H. Wood* as of May 3, 1995, showing closed dome and twin Ross pop valves. Of note is the steam turret with whistle and pipework which leads to injectors on tank tops. Also visible is the rolled-up canvas on the cab roof. The re-railing jack is seen on the far tank and sandboxes on the boiler sides. This is the standard position for the fittings on all IMR locomotives in rebuilt condition with Ross valves and closed domes, although the shape of the pipes do vary on each locomotive.

Sanding gear: Engines Nos. 1-6 had two sandboxes inside the front corner of the cab. The gravity-feed pipe passed round the outside of the coupling rods, dropping sand in-between the two driving wheels. Engines Nos. 7 & 14 were built with two sandboxes just forward of the tanks and two in either corner of the coal bunker, these were mechanically operated by rods. Subsequent engines were supplied with steam sanding gear and all the earlier engines Nos. 1-6 were later modified to this system.

Side tanks: Various size tanks have been fitted to the engines over the years – varying from the earliest small 320 gallon tanks originally supplied with Nos. 1-3, up to the large 520 gallon tanks fitted to No. 16. The original side tanks were of a riveted construction, but in recent years they have been replaced with all welded tanks – only on the engines currently in service.
* See the table on page 25 for a detailed list of individual engine side tank changes.

Side tanks patches: During the decline of the railway during the 1950s/1960s, economy was the watch-word – many of the locomotives had their side tanks repaired with patches instead of being replaced. Most engines received patches in the late 1940s/early 1950s – exception of Nos. 16 and possibly No. 2 which were dismantled in 1951 and No. 7 which was dismantled in 1945. The patches were usually a strip about 8" wide, riveted to lower tank sides – each engine had a different patten of rivets – check photographs of individual engines

Smokebox: The smaller engines, Nos. 1-9 & 14, were built with slopping smokeboxes – Nos. 10-13 & 16 were supplied with the normal type smokebox. The slopping smokeboxes on engines Nos. 4-6 & 8 were replaced with the more conventional type when they received larger boilers. Engines Nos. 1-3, 7, 9, & 14 retained their slopping smokeboxes for their entire life in service.

Smokebox door handles: Engines Nos. 1-7 & 14 were originally supplied with the usual double-handled type. From No. 8 onwards, one handle and a small handwheel were fitted – the earlier engines were also altered to this arrangement as their boilers were replaced.

Steam heating: Introduced in 1937, all engines except No. 7 fitted – fell into disuse as services declined in later years. Never a complete success – see vacuum pipes below.

Toolbox: Photographic evidence shows that engines Nos. 1-3 were fitted with a wooden toolbox mounted on top of the left-hand side tank – the toolboxes seem to have been removed when the engines received larger boilers in the early 1900s.

Vacuum pipes: Continuous brakes were fitted to all engines after an accident in Douglas station in 1923, with the exception of No. 7. Fitting took place over a number of years, but was never a complete success – due to the engines' boilers not being able to cope with extra steam required to operate the vacuum brakes and steam heating. No. 16 was supplied with vacuum brakes already fitted. In more recent times, the continuous brakes are now used on every train as the engines have larger boilers and the loads are lighter.

Wakefield lubricators: A mechanical lubricator mounted on the left-hand running plate, it replaced the original displacement lubricators on the larger 10-13 type and the enlarged Nos. 4-6 type – No. 16 was supplied with it already fitted.

Wheel-balance weights: Locomotives Nos. 1-9 & 14 were supplied with square-ended balance weights. Locomotives Nos. 10-13 & 16 had crescent shaped balance weights.

Wheelbase: On engines Nos. 1-9 & 14 the driving wheelbase is 6'-3", on engines No.10-13 & 16 the driving wheelbase was extended to 6'-6".

Wheel - Number of spokes: Engines Nos.1-9 & 14 the main driving wheels had 11 spokes, later engines having 12 spokes.

Whistle: Locomotives Nos. 1-9 were fitted with a small shrill whistle, later engines having a larger deeper tone one. In recent years, an American chime type whistle has been attached to various engines on special occasions such as the last train of the season, or Santa specials etc.

IMR PROPOSED LOCOMOTIVES

In 1945 the story of the IMR locomotive stud was at the brink of taking a unusual twist. After the end of the Second World War, the IMR was, like most mainland railways, in a very poor run-down condition. To return the railway its pre-war glory, the IMR management started a programme of re-freshing and updating the system ready for 'business as usual'. However, the expected post-war tourist boom of the 1950s never materialized and many of the plans to update the railway had to be shelved.

One of these plans was to purchase new, more powerful locomotives to replace some of the ageing Victorian engines that were, at the time, in bad state of repair. A second scheme was to re-build and enlarge several of the later engines. The current image and charm of the IMR could have been very much different had

No.11 *Maitland* showing clack valve, inside motion, narrow footplate over cylinders and valve gear, with oil pots on top. Just above the main wheel is the Wakefield lubricator (this side only).

No.10 *G. H. Wood* with cast replacement chimney without numerals, smokebox door hand wheel, wooden buffer beam and a collection of oil cans.

either or both schemes ever come to fruition.

The first new type of proposed locomotive from Beyer, Peacock was to be a for a 40-Ton 2-6-2T engine of massive size, power and performance. With an overall length of 31' 10" and tractive effort of 12,640 lbs, this huge locomotive would be capable of handling any train the IMR could muster. The design was an impressive example of narrow gauge motive power, by any standards. However, it was rejected by the IMR as being far too large for its loading gauge. A second modified scaled-down version was then submitted by Beyer, Peacock, which weighed in at 35-Tons with a total length of 28' 6" and a reduced tractive effort of 10,770lbs. For the princely sum of £12,250, this handsome beast and her two proposed sisters would certainly have changed the face of the IMR.

Also under discussion during the same period was the proposal to rebuild engines Nos. 10 to No. 13 and No. 16 into 2-4-2 tank engines. The plans submitted by Beyer, Peacock included extending the frames and adding a new coal bunker of 40 cubic feet capacity and an addition 300 gallons water tank. In order to support this extra weight at the rear of the engine, a new trailing wheelset with radial axleboxes would be incorporated. These modifications would allow the locomotive to run the full 25 miles between Douglas and Ramsey without having to take on additional water. However, the cost of the conversion far outweighed the savings in time and effort.

For the modeller, these proposed engines are well worth considering as additional motive power to your IMR locomotive fleet. They would add a bit of variety and would certainly be a conversation piece, as Brian Caton found out when he built a 5.5mm scale live steam model of the modified 2-6-2T and exhibited it at the Manx Model Rail 95 Exhibition. Brian's model is finished in the appropriate livery of 1950s Indian red and has been given the IMR locomotive number 17 – which would be the next one in sequence. To follow the tradition of naming Manx engines, Brian has fitted nameplates dedicated to the General Manager at that time – A. M. Sheard.

The theme of proposed Manx engines could be taken one step further if desired. What if Beyer, Peacock won the contract to build the MNR's third engine *Caledonia* instead of Dübs? The Beyer, Peacock 0-6-0 side tank would have been very similar in outline to the three delivered to the 3' (914mm) gauge Ballymena & Larne Railway in 1877-83. Or do you fancy a bit of modern image in the form of diesel power? The IMR expressed considerable interest in the three CIE Walker Bros Bo-Bo diesel-mechanical locomotives built in 1955 from the former West Clare line when it closed in 1961. However, the IMR and CIE could not come to an agreement over the price and sadly the three engines, then only six years old, were eventually scrapped. With a little imagination it is quite easy to build up an unusual roster of Manx locomotives.

IMR DIESEL RAILCARS

As the Isle of Man slipped in to a tourist slump during the late 1950s, economy was the watch word on the IMR. Late evening and Sunday trains were suspended, while some stations were closed and part of the rolling stock was put into permanent store. Buses began to replaced certain trains.

In a brave effort to fight this recession, the IMR purchased two railcars in 1961. They were the last power units delivered to the County Donegal Railways in 1950 and 1951 respectfully. No. 19 and No. 20 were only just a decade old when the County Donegal Railways closed on the January 1, 1960. The power units were driven by a Gardener 6LW diesel engine and the cabs were built by Walker Bros. of Wigan. The rear articulated carriage sections were fabricated by in the Dundalk works of the ex-Great Northern Railway of Ireland.

The pair were shipped over to the Island, arriving in Douglas on May 7, 1961. Under IMR ownership they were modified, overhauled and retain their original CDR numbers – Nos. 19 and 20.

Unlike the ex-County Donegal Railway, the IMR has no turntables at its terminus stations. Therefore, the pair have to run back-to-back in normal service. To provide extra luggage and parcel space, a couple of 'G' vans, G5 and G19 were fitted with modified couplings and through-piped for vacuum brakes to run between the two railcars. They were to be found operating mainly on the quieter Peel and Ramsey lines.

The two railcars have not been used in service since the early 1970s. Recently they have been used by the Permanent Way department during the winter months on track repair duties, while in the summer months, they are used as station pilots at Douglas station, shunting the coaching stock. During a hot dry summer such as 1995, the pair follow a service train up Port Soderick Bank pushing the fire fighting truck to dampen down any trackside fires started by a stray spark from the steam locomotives.

The small builders' plate of No.11 *Maitland*.

No.4 *Loch* of 1874 enters Ballasalla on May 5, 1993 with the 10.15 train from Port Erin to Douglas. To the untrained eye there is nothing unusual about this photograph, but any IMR enthusiast worth their salt will soon notice that the locomotive is facing the wrong way. Since there are no turntables on the island, all locomotives run chimney–first out of Douglas, but in 1993 No. 4 was turned round and ran for two years in this manner causing quite a stir.

No.9 *Douglas* at Port Erin on September 3, 1993. This 1896 locomotive still has its original sloping smokebox and Salter safety valves, and is presently in store. The livery is not a IMR one. The locomotive appears to be in steam, but the effect was created by burning a tyre in the smokebox.

The remains of No.7 *Tynwald* of 1880 – dismantled in 1945 – at Santon station in 1985. What's left of No.7 is presently on display at Castletown station.

Happy days at Douglas in August 1964 with Nos.8, 5 and 11 lined up with an equally impressive collection of semaphore signals. TREVOR NALL

With the arrival of the new diesel locomotive in 1992 (see below), the two ex-CDR railcars were effectively redundant. The bodywork on both is currently in a very poor state of repair, although mechanically they are sound. As they now prove to be very popular with many visiting enthusiasts, plans are in hand to refurbish these two historical units and return them to service.

Anbrico produced a white metal body kit of the ex-CDR railcars many years ago. A revised version of this kit is due to be produced by Roxey Mouldings, which will include a motorising kit. Just to confuse the modeller – the two vehicles are not identical. There are several detail differences between each unit. The most noticeable is the front cab windows, the small sliding windows in the carriage portions and roof vents – check photographs for details.

IMR DIESEL LOCOMOTIVE

Until 1992, The Isle of Man Railway locomotive fleet consisted entirely of steam locomotives. This presented many operational problems, such as when one of the steam engines failed or, due to increased traffic demands, an extra locomotive would be required. Raising steam on these rather elderly ladies can take anything from two to four hours, depending on conditions. In attempt to reduce operating costs and to provide more flexibility, a second-hand Schoema diesel locomotive was purchased from Germany. Its main duties were to be on works trains during the winter months when it is uneconomical to run a steam locomotive, and during the summer operating season it would be used in emergency situations.

The 'Schoema' had been in store for about four years at a railway works in Dortmund. Built in 1958, originally it had worked on a 900mm gauge mine system in Helmstedt. The loading gauge was similar to that of the IMR and the 900mm gauge was only 14mm less than that of the Island's railways – in theory, it should fit straight on to the tracks!

The 'Schoema' was shipped to the Island on seven days approval, arriving on the July 18, 1992 and the 'foreigner' caused quite a stir when it was unloaded at Douglas station. The engine was, in fact, the first new locomotive on the IMR since 1926, when the last Beyer, Peacock 2-4-0T No. 16 *Mannin* was delivered.

The 24-ton diesel is an 0-4-0, built in 1958, powered by a V12 Deutz air-cooled engine, which is turbo-charged delivering 200hp to a Voight three speed automatic gearbox.

A close look at the cylinders and pony truck of No.4 *Loch*. Note the lining on the pony truck axlebox cover and very oily condition of the pony truck in general due to the cylinder drain cocks being left open spraying steam and oil over the pony truck.

Although not as fast as the snappy steamers, the diesel is very powerful and has been a useful addition to the locomotive roster after a few minor teething troubles. The engine has lost its 'foreign' look after a trip to the IMR's work and paint shops.

Norwegian-style chopper couplings, side chains and vacuum brakes have been fitted. The diesel has been outshopped in a smart livery of lined IMR dark green and given the next number in the locomotive fleet sequence – No. 17. A competition was held in 1993 to find a suitable name for this diesel. The winner was Viking, which was to be the original name of No. 3 *Pender* of 1873, but it was changed at the last minute.

There are no models of No. 17 *Viking* currently available at the time of writing. However, it would not be to difficult to scratch-build a body to fit on to a commercial TT 12mm gauge 0-4-0 diesel chassis. *Viking* is a essential feature of an 'modern image' IMR layout.

THE MANX NORTHERN RAILWAY QUARTET

While the engines of Isle of Man Railway locomotive fleet consisted

The newest locomotive is No.17 *Viking*, a Schoema 4-wheeled 24-ton diesel, built in 1958. It has been fitted with Norwegian chopper couplings and side chains, which are standard on the IMR.

A 1962 view of the ex-County Donegal Railway railcars at Douglas. TREVOR NALL

IMR No.10 *G. H. Wood* of 1905 on shed in between turns at Douglas station. *G. H. Wood* carries the IMR 1873-1945 dark green livery. Apart from the closed dome and Ross 'pop' valves, No.10's appearance has changed very little over the years. The locomotive spent a number of years out of action and was returned to service in 1993 after receiving the 1971 boiler from No. 13 *Kissack*.

Ex-Manx Northern Railway 0-6-0 No.4 *Caledonia* storming up the Snaefell Mountain Railway in wintry conditions on March 29 1995. 'Cale', as it is nicknamed, was used on the SMR during 1995 to help celebrate the line's centenary. The 'Cale' was hired from the MNR in 1895 by the contractors to help build the SMR.

A special night photography evening was held at Douglas station as part of the 1995 International Railway Festival. Lined up are Beyer, Peacock 2-4-0T No.10 *G. H. Wood* and ex-Manx Northern Railway 0-6-0T No.4 *Caledonia*.

Ex-Manx Northern Railway No.4 *Caledonia*, a Dübs-built 0-6-0 of 1885 in MNR Tuscan red livery at Douglas station on August 8 1995. *Caledonia* was returned to service in 1995 after nearly 30 years in store. The Island's only 0-6-0 is legendary among IMR enthusiasts.

Manx Peacock No.4 *Loch* of 1874 on shed at Douglas station during 1994. *Loch* is one of the world's oldest operating steam engine still in service on its original line. No.4 is also famous for its appearances on the 'Steam on the Manx Electric Railway' trips.

Brian Caton's 5.5mm scratch-built live steam model of the proposed, but never built, Beyer, Peacock 2-6-2T No.17 *A. M. Sheard* in 1950s Indian red livery.

exclusively of a standardized design – all from the same builder, the Manx Northern Railway on the other hand owned three different designs of engines from three different builders. The main reason for this large diversity of motive power was a financial one. The MNR was short of cash from day one, so the choice of rolling stock was down to the matter of price. The original plan was to purchase three locomotives for the opening of the line in 1879. However, only two were ordered as there was not enough available cash for the third.

Sharp, Stewart of Manchester, with the lowest tender, won the contract in 1879 to supply two 2-4-0Ts. Three sightly smaller versions of the Manx Sharpies were supplied to the 3' (914mm) gauge Southwold Railway in the same year.

The 'Sharpies' were built to a very similar specification to that of the Manx 'Peacocks'. The driving wheel diameter and cylinders were identical, while the Sharp, Stewarts had slightly larger sidetanks of 400 gallon capacity and a bigger firebox. The general outline of the 'Sharpies' was more of an orthodox design compared with the ornate Manx 'Peacocks'. One major difference was that the leading wheel on the 'Sharpies' was mounted in a radial box, while the Beyer, Peacocks had the American Bissel type pony truck – the first locomotives in the British Isles to be so fitted. The 'Sharpies' radial boxes were to be a constant source of trouble throughout their lives. The engines hunted badly at speed and tended to spread the gauge on the curves. They were named MNR No. 1 *Ramsey* and MNR No. 2 *Northern.*

A third engine was desperately need by 1880. Yet again Sharp, Stewart offered the cheapest price, however they did not like the MNR £5 per day penalty clause for late delivery. So, the contract was

An impressive 5" scale live steam model of No.5 *Mona* of 1874.

No.5 *Mona* of 1874 stored in the back of Douglas carriage shed, where it has been for more than 25 years. Hopefully, one day it may be restored and returned to service.

awarded to Beyer, Peacock. The new engine was a slightly modified version of the Manx Peacock 2-4-0T design. MNR No. 3 was built at Beyer's at the same time as IMR No. 7 Tynwald of 1880. The engine was delivered to the MNR on time and was named *Thornhill* after the MNR Chairman, J. T. Clucas's house.

By 1885, a fourth engine was necessary to meet the increasing traffic requirements due to the opening of the Foxdale Railway, which was operate by the MNR. Against this background was born one of the biggest legends in Manx railway history – *Caledonia*.

Caledonia – or 'Cale' as it is known locally – is the largest, heaviest, and most powerful steam locomotive ever to operate on the Isle of Man's 3' (914mm) narrow gauge systems and being the Island's only 0-6-0T engine is the odd man out, all the other IMR and MNR engines were of a 2-4-0T design.

Purchased by the MNR, *Caledonia*'s original duties were to haul the heavy lead ore trains from the Foxdale mines down to the quayside at Ramsey harbour, and for the return journey back again with coal for the mines pumping engines. It was built by Dübs of Glasgow and arrived on the Island in late December 1885.

The locomotive entered service in January 1886 as MNR No. 4, the engine was used mainly on the Foxdale branch until 1911, when the mines closed. *Caledonia* was hired from the MNR in 1895 to help construct the electric railway up Snaefell mountain. It was used to push loaded contractors' wagons up the steep 1 in 12 gradient to the summit, although a third rail had to be laid to allow the 3' (914mm) gauge 'Cale' to operate on the 3' 6" (1,066mm) SMR.

For most of the locomotive's life under IMR ownership, *Caledonia* was banished to the back of Douglas engine shed, only ever emerging for cattle show specials or snow clearance duties. As its tractive effort of 11,122 lb was almost double that of any of the other Island's engines, it was very useful for these duties and saved the need for costly double-heading.

Caledonia last ran on June 2, 1968 on an Isle of Man Steam

STEAM LOCOMOTIVES: Isle of Man Railway Co.

No.	Name	Year	Builder	No.	Type	Notes	Remarks
No. 1	*Sutherland*	1873	Beyer, Peacock	1253	2-4-0T	A, E	On Display
No. 2	*Derby*	1873	Beyer, Peacock	1254	2-4-0T	A, E	Scrapped 1951
No. 3	*Pender*	1873	Beyer, Peacock	1255	2-4-0T	A, E	On Display
No. 4	*Loch*	1874	Beyer, Peacock	1416	2-4-0T	B, F	In Traffic
No. 5	*Mona*	1874	Beyer, Peacock	1417	2-4-0T	B, F	In Store
No. 6	*Peveril*	1875	Beyer, Peacock	1524	2-4-0T	B, F	On Display
No. 7	*Tynwald*	1880	Beyer, Peacock	2038	2-4-0T	B, E	Dismantled 1945
No. 8	*Fenella*	1894	Beyer, Peacock	3610	2-4-0T	B, E, H	Being Rebuilt
No. 9	*Douglas*	1896	Beyer, Peacock	3815	2-4-0T	B, E	In Store
No. 10	*G. H. Wood*	1905	Beyer, Peacock	4662	2-4-0T	C	In Traffic
No. 11	*Maitland*	1905	Beyer, Peacock	4663	2-4-0T	C, G	In Traffic
No. 12	*Hutchinson*	1908	Beyer, Peacock	5126	2-4-0T	C, G	In Traffic
No. 13	*Kissack*	1910	Beyer, Peacock	5382	2-4-0T	C	In Store
No. 14	*Thornhill*	1880	Beyer, Peacock	2028	2-4-0T	B, E, J	Privately Owned
No. 15	*Caledonia*	1885	Dübs	2178	0-6-0T	K	In Traffic
No. 16	*Mannin*	1926	Beyer, Peacock	6296	2-4-0T	D	On Display

STEAM LOCOMOTIVES: Manx Northern Railway Co. (up to 1905)

No.	Name	Year	Builder	No.	Type	Notes	Remarks
No. 1	*Ramsey*	1879	Sharp, Stewart	2885	2-4-0T	L	Scrapped 1923
No. 2	*Northern*	1879	Sharp, Stewart	2886	2-4-0T	L	Scrapped 1912
No. 3	*Thornhill*	1880	Beyer, Peacock	2028	2-4-0T	B, E, J	Privately Owned
No. 4	*Caledonia*	1885	Dübs	2178	0-6-0T	K	In Traffic

DIESEL LOCOMOTIVES: Isle of Man Railway Co.

No.	Name	Year	Builder	No.	Type	Notes	Remarks
No. 17	*Viking*	1958	Christoph Schottle	2175	0-4-0	M	Per. Way Dept.
No. 18	-	-	-	-	-	-	-
No. 19	-	1950	Walker/GNR	-	Railcar	N	Per. Way Dept.
No. 20	-	1951	Walker/GNR	-	Railcar	N	Per. Way Dept.

Notes:

A: Built with a small 2'-10 3/4" dia. boiler and 320 gallon tanks.
B: Built with a small 2'-10 3/4" dia. boiler and 385 gallon tanks.
C: Built with a medium 3'-3" dia. boiler and 480 gallon tanks.
D: Built with a large 3'-6" dia. boiler and 520 gallon tanks.
E: Rebuilt with a 3'-3" dia. boiler and 385 gallon tanks.
F: Rebuilt with a 3'-3" dia. boiler and 480 gallon tanks.
G: Rebuilt in 1981 with a large 3'-7" dia. boiler.
H: Re-entered service in 1993 with ex-No.13's boiler.
J: Ex-Manx Northern Ry. engine: No.14 ex-MNR No. 3.
K: Ex-Manx Northern Ry. engine: No.15 ex-MNR No. 4.
L: Manx Northern Ry. engines: No.1 & No. 2 – seldom used after 1905 by IMR.
M: Purchased from a 900mm gauge mine system in Helmstedt, Germany in 1992.
N: Purchased from the former County Donegal Railway in 1961.

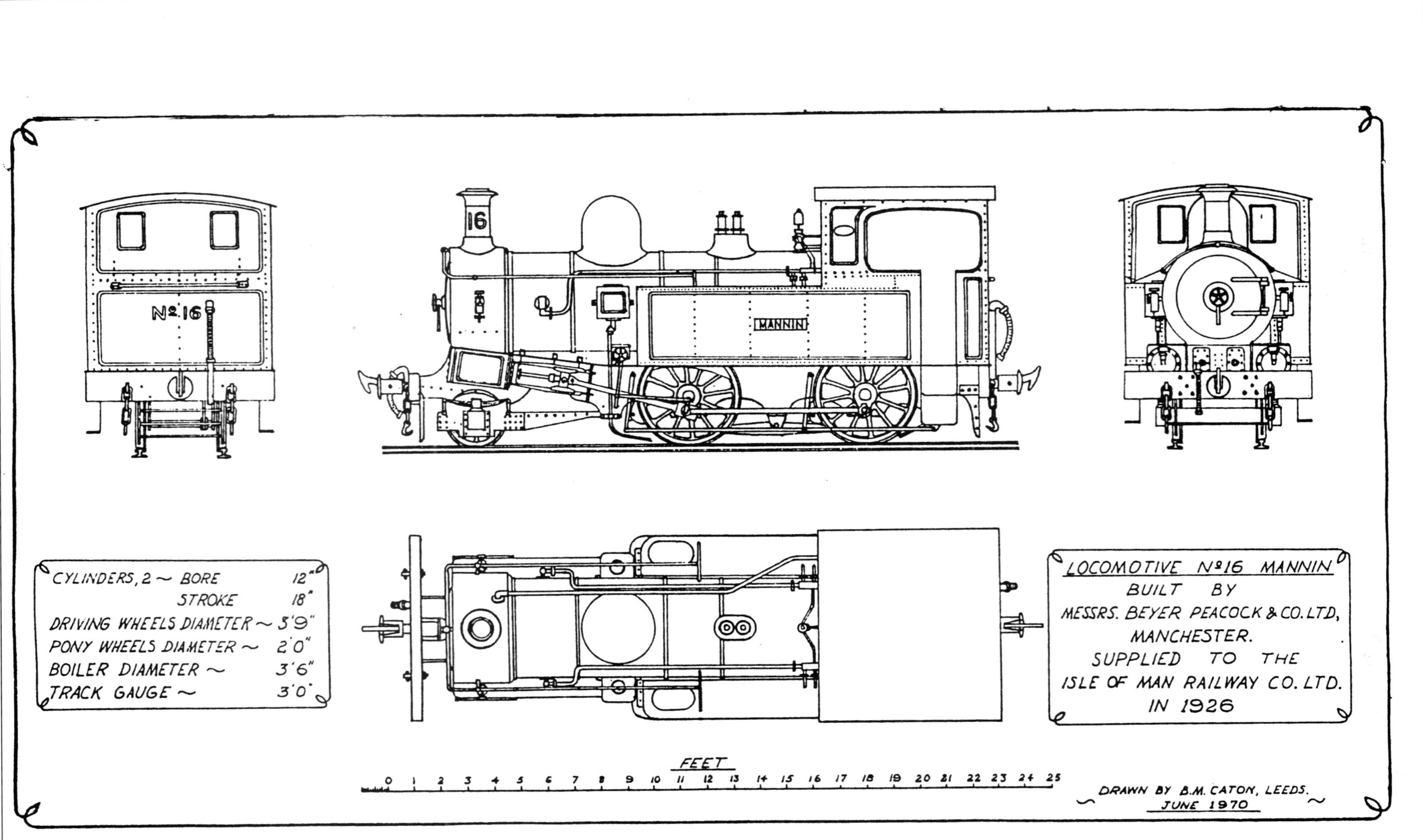
No 16
16
MANNIN
CYLINDERS, 2 ~ BORE 12"
STROKE 18"
DRIVING WHEELS DIAMETER ~ 3'9"
PONY WHEELS DIAMETER ~ 2'0"
BOILER DIAMETER ~ 3'6"
TRACK GAUGE ~ 3'0"
LOCOMOTIVE Nº16 MANNIN
BUILT BY
MESSRS. BEYER PEACOCK & CO. LTD,
MANCHESTER.
SUPPLIED TO THE
ISLE OF MAN RAILWAY CO. LTD.
IN 1926
FEET
0 1 2 3 4 5 6 7 8 9 10 11 12 13 14 15 16 17 18 19 20 21 22 23 24 25
~ DRAWN BY B.M. CATON, LEEDS. ~
JUNE 1970

BOILER CHANGES:

No.#1 =1891, new boiler
1919, No.7's of 1903
1923, No.5's of 1907
1939, No.7's of 1923#

No.#2 =1893, new boiler
1912, No.6's of 1892
1917, No.2's of 1893
1921, No.5's of 1895
1923, new boiler#

No.#3 =1888, new boiler
1913, new boiler
1951, No.2's of 1923

No.#4 =1893, new boiler
1909, new boiler
1946, No.5's of 1914
1968, new boiler*

No.#5 = 1895, new boiler
1907, new boiler
1914, new boiler
1936, No.6's of 1911,
1946, new boiler*

No.#6 =1892, new boiler
1911, new boiler
1932, new boiler*

No.#7 =1903, new boiler
1919, No.5's of 1907
1923, new boiler#
1939, No.5's of 1907

No.#8 =1908, No.5's of 1895
1915, No.5's of 1907
1919, new boiler
1936, new boiler@

No.#9 =1909, No.8's of 1894
1912, new boiler

No.#10 =1926, new boiler*
1948, new boiler*
1993, No.13's of 1971)

No.#11 =1934, new boiler*
1959, new boiler*
1980, new boiler*

No.#12 =1946, new boiler*
1980, new boiler*

No.#13 =1944, No.11's of 1905
1949, No.10's of 1926
1971, new boiler*

No.#14 =1896, new boiler
1910, No.4's of 1893
1913, No.2's of 1893
1916, No.5's of 1895
1921, new boiler

No.15 1922, new boiler

No.16 None.

Notes:

* Fitted with a closed dome and Ross`pop' valves.

Fitted with Ross `pop' valves one on top of the boiler, the other housed in the top of the dome.

@ Special 2'-10¾" boiler with a closed dome and Ross `pop' valves

Supporters Association special. In 1975, the locomotive was moved to a more permanent home on display in the Port Erin railway museum, where it remained until September 23, 1993.

As nearly every Isle of Man Railway enthusiast everywhere had dreamt of seeing *Caledonia* back in action again, so the Isle of Man Railway thought it would be fitting to overhaul 'Cale' and run it up the mountain to help celebrate the centenary of the Snaefell Mountain Railway in 1995.

After a major overhaul, *Caledonia* was back in action delighting the crowds during 1995. Not only was 'Cale' seen operating on her home ground of the IMR, but ran on the upper section of the Snaefell Mountain Railway and the Manx Electric Railway. During 1996, *Caledonia* is due to operate regular service trains once again – the legend grows!

The four MNR engines were taken into the IMR locomotive fleet when the MNR was absorbed in 1905. *Thornhill* and *Caledonia* were allocated IMR No. 14 and No. 15 respectfully, while the two 'Sharpies' were unofficially given Nos.16 and 17. As the two 'Sharpies' were non-standard, in a poor condition and not very popular with IMR staff, sadly they were scrapped. Ex-MNR No. 1 *Ramsey* was scrapped in 1923, while ex-MNR No. 2 *Northern* was lost a number of years earlier in 1912. IMR No. 14 *Thornhill* finally succumbed in 1963 and was sold several years later for private preservation on the Island.

The Manx 'little and large'– In October 1995, IMR No.11 *Maitland* and Groudle Glen Railway's *Sea Lion* are seen in Douglas workshops for boiler and re-tubing repairs.

On May 1 1995, No.12 *Hutchinson* of 1908, awaits the arrival of the Up train in Ballasalla's passing loop.

After the Manx 'Peacocks', the second most requested Manx engine by modellers is *Caledonia*. A 4mm scale cast resin kit of the 'Cale' has been available for several years from Malcolm Mills. This was joined in 1995 by a highly detailed white metal and etched nickel silver *Caledonia* kit from Branchlines.

Branchlines also produce a small-boilered Beyer, Peacock kit, which can be built to represent *Thornhill* in either MNR and later IMR days. With the current popularity of the Island's railways, it can only be a matter of time before the two MNR 'Sharpies' are produced in kit form in 4mm scale. It will then be possible to construct the entire MNR locomotive fleet.

SUPER-DETAILING YOUR MANX LOCOMOTIVES

This section is aimed at improving the 4mm scale locomotives, as at present this is the most popular scale with Manx modellers. However, a lot of these super-detailing notes could apply equally to builders of the Manx 'Peacocks' in other scales.

As supplied, Branchline's 4mm kit of the IMR Beyer, Peacocks 2-4-0T is very well detailed. These detailing parts are also available separately to super-detail the old Gem locomotive kits.

Many modellers will be content to attach their favourite nameplates. However, still further details can be added by studying the notes in this book and photographs of individual engines. This should enable the builder to produce a first class, highly super-detailed model of one of the Manx 'Peacocks'.

Starting first with the bodywork. To greatly improve running qualities, pack as much lead or similar into the boiler, tanksides, coal bunker and under the cab roof. These little engines need as much weight as you can possibly cram in. The coal bunker needs finishing with a topping of real crushed coal and do not forget to have a pile of coal on the cab floor. The footplate crew normally have to stand on pile of coal for the first half of the trip as there is precious little space inside the cab of a Manx 'Peacock'. Some engines have little wooden boards that slot in between the bottom of the cab handrails that allows the fireman to carry even more coal on the cab floor!

The cabside canvas draught dodgers are an important feature of these locomotives. If you study various photographs of these engines, you will notice that they are in one of three positions.

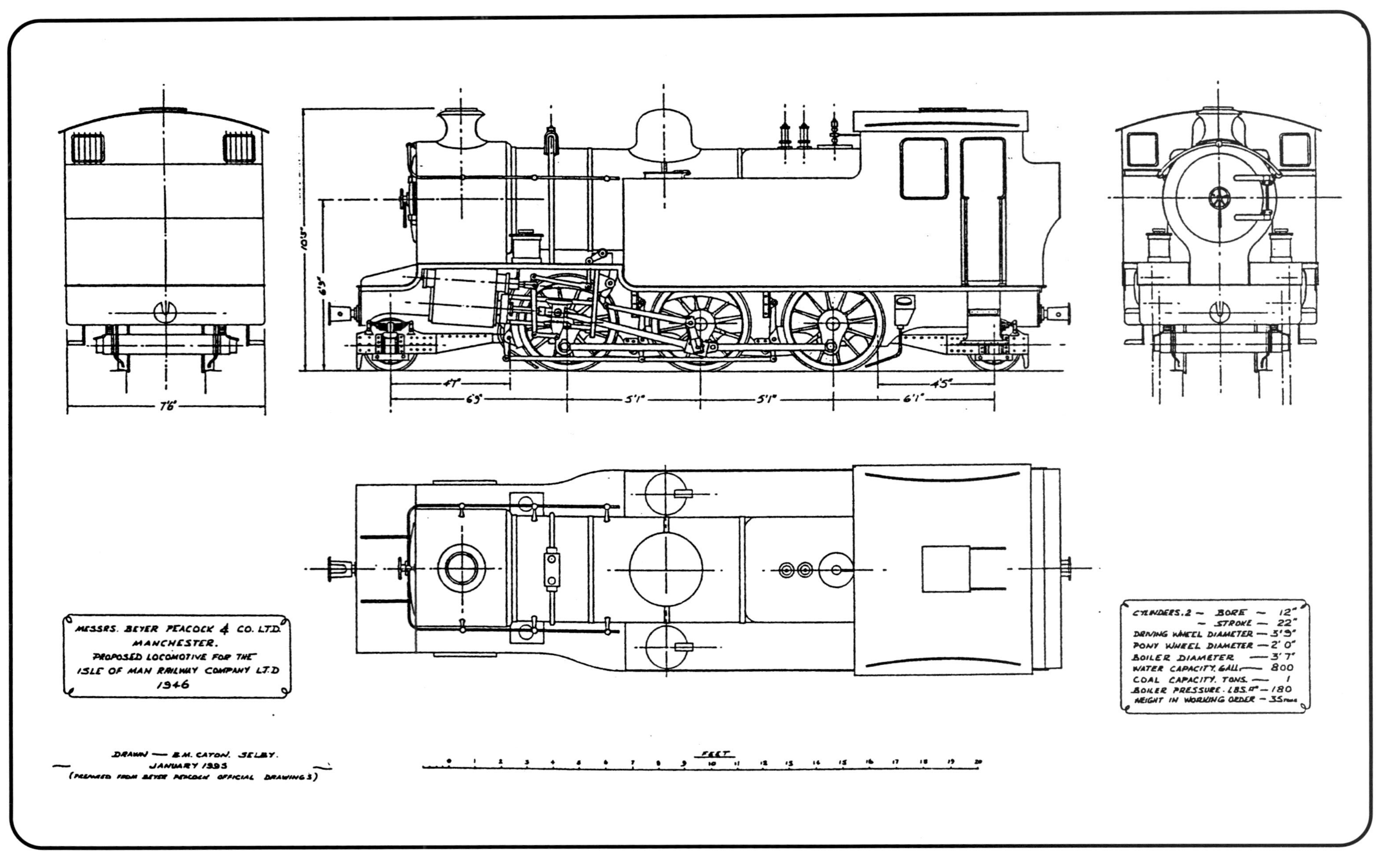
CYLINDERS. 2 ~ BORE ~ 12"
~ STROKE — 22"
DRIVING WHEEL DIAMETER — 3'9"
PONY WHEEL DIAMETER — 2' 0"
BOILER DIAMETER — 3' 7"
WATER CAPACITY. GALL. — 800
COAL CAPACITY. TONS. — 1
BOILER PRESSURE. LBS.□" — 180
WEIGHT IN WORKING ORDER — 35 TONS
MESSRS. BEYER PEACOCK & CO. LTD.
MANCHESTER.
PROPOSED LOCOMOTIVE FOR THE
ISLE OF MAN RAILWAY COMPANY LTD
1946
DRAWN — B.M. CATON. SELBY.
JANUARY 1995
(PREPARED FROM BEYER PEACOCK OFFICIAL DRAWINGS)
FEET
0 1 2 3 4 5 6 7 8 9 10 11 12 13 14 15 16 17 18 19 20
7'6"
10'5"
6'9"
4'7"
6'9"
5'1"
5'1"
6'1"
4'5"

A busy scene at Douglas engine shed in July 1995. No.10 *G. H. Wood* takes on coal ready for its next duty. An ex-Manx Northern Railway six-wheeled coach body is used as crew mess hut. Other points of interest include a vertical boiler to drive the overhead belt shaft in the workshops, and a vertical belt-driven drill ready to be installed. The road trailer was for transporting the 2' gauge Groudle Glen Railway locomotive *Sea Lion* to the IMR workshops.

Either rolled up on the cab roof – which is the current practice, unrolled fully open for really bad weather, or the most common position is unrolled, but pushed forward and tucked into the small cabside handrail. The best way to tackle them in 4mm scale is find a section out of an old magazine that is the correct colour – the canvas always seemed a very grimy grey-brown colour. Now screw the paper up several times to produce the well-creased look, cut to shape and glue on. A couple of coats of matt varnish finishes the job off nicely. While your are up on the cab roof, this needs painting black or grimy dark grey – many modellers seem to fall into the trap of painting their Manx 'Peacocks' roof the main body colour.

The distinctive sidetank patches that appear on many IMR locomotives in later years can be made from strips of five thou. plastic card embossed with rivets. However, be careful as each engine had a different patten of rivets! The narrow footplate that runs along the bottom of the sidetanks on both the Branchlines and Gem kits are a little thick due to the whitemetal casting process. By filing a slight chamfer to reduce the thickness on the footplate edge vastly improves the locomotive's appearance. The addition of three little brass oiling pots on the inclined section of either footplate over the slide bars can be made from 1.5mm length of 1mm diameter plastic rod.

The wooden bufferbeams are crying out for extra detail, especially the front one, which is very prominent. I have replaced the front cast bufferbeam with a 2mm plywood version with the characteristic split along the grain that was a feature of many of the Manx engines.

Both bufferbeams need a coupling plate made of five thou. plastic card and large rivet detail. These rivets can be made from small drops of superglue gel applied with the point of a cocktail stick. It takes a little practice to get the right size drop on the stick, but the results are well worth the effort.

Also missing off the bufferbeams are the lampirons. The rear pair are a bit unusual as they actually go through a slot in the footplate, while the two front pair of lampirons can be either fitted to the front or rear of the beam depending in which period your particular engine you are modelling is based – it is best to check photographs first. The front faces of the bufferbeams are painted vermilion, while the rear faces are black. The ends of the bufferbeams are painted the same as the body colour complete with lining!

Between the chassis frames, particularly on the smaller boiler type, is quite bare and with the inside of the frames being painted vermilion, it draws attention to this void. This area really requires some dummy inside motion to fill the empty space. I find that a couple of non-working coupling rods and the middle frame spacer which these pass through is usually adequate. However, if you really want fully working inside-motion there is plenty of space to fit it all in – if you like that sort of thing of course!

Also, quite a lot of work can be done to improve the pony truck supplied with the Branchlines kits. As supplied it is a simple fold-up affair, but needs filling in with extra detail. The top section of the pony truck is missing. This can be covered in with a sheet of plastic card or brass. Plastic card is probably better to cut down the risk of unwanted short circuits in this area. In fact, as there is very little clearance between the top of the pony truck and bottom of the cylinders. This can cause electrical problems when the pony truck swings out on curves. The replacement of the brass tube used for the cylinders on the Branchlines kit with a plastic tube of the same diameter eliminates these problems.

As there is loads of oily steam sprayed out of the cylinder drain cocks, the pony truck need to be painted a very oily black. Do not forget the pony truck's axlebox covers are also painted the main body colour, complete with lining.

The rear stone guards need three round bars between them, which transforms it into a mini cow-catcher for cab first running. Also, while you are in the mood for detailing, the tank balance pipe that joins the two sidetanks can be added if you wish. However, this is normally well out of sight behind the cab steps. The tank filler caps are situated inside the cab on the tank tops. These are just a 6" round wooden plug, which can be represented by a 2mm diameter by .75mm thick disc of plastic card.

The whistle on the Branchlines kits needs altering. First, it requires a rounded top and, if it is one of the early Nos.1-9 or No.14, it needs a shorter whistle. The locomotive lining is quite complex and very fine, unless you are a dab-hand with a lining pen, N gauge lining is just the right size for the 4mm scale models. British Railways 1950s coach yellow/black/yellow lining is perfect match for the lining on the IMR Indian red engines.

The finishing touch to your 4mm Manx engine is the name and builders' plates. Branchlines have introduced an etched brass fret containing all 15 IMR and the four MNR nameplates, builders plates and chimney numerals to complete your model. These small detail improvements, which are not too difficult or too costly in time and money for the average modeller, will make your model of the Manx 'Peacock' stand out in the crowd.

The 4mm scale ex-MNR *Caledonia* requires very little extra detailing. Just the usual footplate crew, coal in the bunker and nameplates to complete. A nice addition would be the large snow plough carried during the later years with the IMR.

Chapter Four

IMR & MNR CARRIAGES

F18 at Ballasalla station on May 14 1995, as part of the current service fleet. The livery is purple lake with white uppers. Note the double footboards – essential for the stations with no platforms!

By narrow gauge standards, the 3' gauge Isle of Man Railway once owned a huge fleet of passenger carriages, in fact this Manx railway had even more coaches than some mainland standard gauge companies! The main source of revenue for this Manx railway has always been passenger traffic and at its peak, the railway possessed 75 bogie-carriages and 14 six-wheeled coaches to cope with the huge annual influx of tourists to the Island during the summer season, where double-headed trains of up to 14 coaches would be an everyday sight. However, this was offset by the poor winter service, where only one or two coaches and a couple of wagons was all that was needed to operate during the long dark winter months. The IMR started out by purchasing a number of four-wheeled coaches for the opening of the Douglas–Peel line in 1873, and Douglas-Port Erin line the following year.

The four-wheelers were obtained as a part of an economy measure, as the railway was desperately short of cash due to the difficulty in raising the necessary capital. A total of 54 four-wheeled coaches were supplied by Metro-politan Carriage & Wagon Co., Saltley, but they soon proved to be totally inadequate, as the railway was far more popular than originally anticipated. The extra traffic was due to the ever increasing numbers of Victorian and Edwardian visitors arriving on the Island each year, putting a strain on the railway's resources. To combat the increasing heavy loads, The Isle of Man Railway Co. in 1876 took delivery of the first batch of larger bogie-coaches from Brown Marshalls. Prior to 1876, the IMR had numbered both their coaches and wagons as it received them, from No. 1 upwards, which was the normal practice on small railways. After 1876, a new system of identification was introduced, were each different type of coach, brake van and wagon had a prefixed letter – this system still remains in use today. Starting with the original 54 small four-wheeled stock:

Series A = A1 to A12 – The four-wheeled, three compt. all 1st Series
B = B1 to B24 – The four-wheeled, three compt. all 2nd
Series C = C1 to C14 – The four-wheeled, three compt. 2nd/Guard
Series D = D1 to D2 – The four-wheeled, three compt. 1st/2nd comp
Series E = E1 to E10 – The four-wheeled luggage/brake vans

In order to reduce marshalling and shunting time, from 1887 these four-wheeled coaches were close-coupled together in 26 pairs – these consisted of a balanced mixture of 1sts and 3rds. In later years, between 1909 and 1926, the 26 closed-coupled 'pairs' bodies were mounted on a new single metal bogie chassis and were given 'F' series Nos F50 to F75 some time between 1927 and 1933 – they being:- B7 + B8(F50) B3 + B5(F51) A2 + C2(F52) A5 + B21(F53) A7 + C10(F54) B2 + C6(F55) A8 + C8(F56) B16 + B20(F57) B18 + C3(F58) A6 + C4(F59) B13 + B24(F60) A10 + C12(F61) A1 + B1(F62) B6 + B10(F63) B19 + C1(F64) B22 + C7(F65) B11 + B15(F66) B23 + C14(F67) A9 + C13(F68) B4 + B17(F69) B9 + B14(F70) B12 + C5(F71) A3 + D2(F72) A4 + D1(F73) A11 + C11(F74) A12 + C9(F75) *Later 'F' series Nos. in brackets.

Next, all the new IMR bogie carriages as they arrived were allocated an 'F' prefix, this series eventually totalled 75 coaches and is the most complicated series of all the IMR rolling stock (See IMR Carriage list on page 35 for full details of the IMR and ex- MNR passenger carrying stock).

Passenger train formation: Basically anything goes! – Trains are made up according to the time of year, a particular day of the week, a particular time of day and, most importantly, by the weather (it is said that the IMR receipts rise and fall with the barometer).

Therefore, during the busy summer season all coaches would be

IMR saloon F31, built by Metropolitan in 1905, at Douglas station, July 1995, as part of a three-coach 'Bar Set' with F35/36. Short footboards are fitted beneath the doors only. A 4mm scale kit for this vehicle is produced by Roxey.

Large 'F' F48 built by Metroplitan in 1923. The next coach is a small 'F', demonstrating the height difference in the two types. The location is Douglas station, July 1995. Branchlines produce a 4mm scale kit for this coach type.

Small 'F' F15 of 1894. The seating is all 3rd – despite what it says on some of the doors! The white roofs indicate a recent paint job – soon it will be sooty grey! A Roxey kit is available for these coaches.

Small 'F' F11 built by Brown Marshalls in 1881, the subject of a 4mm Roxey kit. Seen here at Douglas in July 1995, it had been recently outshopped with a white roof.

required to cope with the very heavy loads, while in the winter months, a vast majortiy of the coaching stock would be put into hibenation in two huge carriage sheds at Douglas and St Johns, and a smaller one at Ramsey. The winter service trains could consist of only one 'Half-Brake' (a G–/3/3/3 type) coach, which would be strenghed by one or two extra 'Small F' coaches on the slightly busier trains. Also operating during the winter, was the infamous school specials bringing school children from Peel and the surrounding countryside to – and from the secondary schools in Douglas and Ramsey – the story of the kids antics on these trains was at one stage even reported in the 'News of World' newspaper! A rake of rather worn-out 'Pairs' coaches in the overall dark red-brown livery was used for this twice daily Manx 'St Trinans' special, which was eventually replaced by a bus service in the early 1960s, much to the relief of the railway staff!

The losses in revenue from the poor winter service was recuporated by the short hectic summer season, where the IMR could hardly cope with huge crowds of holiday makers. In its heyday in the 1920s, the Isle of Man Railway operated 100 trains daily, carrying a million and a half passengers each year. The most profitable of the lines was the busy southern line to Port Erin, which would carry nearly twice the passengers than the two other lines to Peel in the west and Ramsey in the north. During the summer season the carriages would be marshalled in sets, usually three or four coaches with one of the 'half-brake' (G- /3/3/3) type normally placed at the Douglas-end of the train nearest to the exit and the boat. These distinctive coaches provided essential accommodation for holidaymakers, luggage and small goods.

At the other end of the set would be another brake coach – a five compartment brake, with the other coaches in between being the non-braked type. i.e. F45/F47/F42 or F46/F47/F48/F49. To these basic sets, additional carriages would be added as traffic demanded, often making the early morning trains out of Douglas and the returning late afternoon trains from Port Erin very long, 13 and 14 coach trains were very common during the summer months. i.e. locomotive No.12 & locomotive No.16 + F35/F50/F61/F5/F29/ F1/F10/F18/F56/F45/F48/F49. Just to confuse matters even more, often the Ramsey & Peel trains – departing from Douglas would be combined into one, with the Peel section normally at the rear. Once at St John's, the Manx Crewe junction, the train would be split into the two sections just before entering the station itself. Both trains to Ramsey and Peel would then often leave St John's at the same time on parallel tracks, where an unofficial race would take place – the Peel train always won as it was on a falling gradient! However, there does not seem to set pattern of coach formations on these two lines, although, the older bogie stock seems to be used more on the lighter passenger traffic of the Ramsey and Peel lines. However, after many hours of study, there does not seem to be any obvious pattern as to how these trains were strengthened with extra coaches' as I said before – anything goes. My best advice is to obtain one or two of the videos featuring vintage footage of the Island's railways and copy some of the coach formations on these – at least you will know what you are running on your layout will be accurate. Three excellent videos worthy of note are the Oakwood Video Library No. 3 'Manx Lines Through the Years' and Railscene's to The Ivo Peters Collection Nos. 9 and No. 14'.

Bogies:

There are two types of bogies fitted to IMR coaches. The early carriages up to F44 and oddly F49 were supplied with 4' 6" wheelbased diamond framed bogies with 2' 3" diameter wheels. Later coaches F45 to F48 and the 'Pairs' chassis F50 to F75 were delivered with pressed-steel bogies with a 4' 9" wheelbase and with 2' 3" diameter wheels. F9 to F11 were fitted with pressed-steel bogies of scrapped 'Pairs' coaches in the early 1980s.

Coach lighting:

The original oil lamps on both the IOMR & MNR carriages were soon converted to electric lighting to compete with the modern Manx Electric Railway which opened in 1893. All the 'F' series stock had been converted by 1911, with the exception of the two 'Empress Vans' (F27 & F28), which were not tackled until 1924 and the 'Foxdale Coach' (F39) in 1926. Some of the MNR six-wheelers lost their electric lights between 1910-12 and had their oil lamps

'Pairs' F63 built from old four-wheelers B6-B10 in 1920, still extant in Douglas carriage shed. Notice the 9" strip between bodies, which was originally left open – for cleaning?

IMR 'F' SERIES CARRIAGE LIST:

Number	Date	Makers	Type	Notes	Status	4mm Kit
F1	1876	Brown Marshalls	G/3333/3	A	Lost - Fire 1975	Roxey 4NGC5
F2	"	" "	"	A	Scrapped - 1976	"
F3	"	" "	"	A	Scrapped - 1975	"
F4	"	" "	"	A	Lost - Fire 1975	"
F5	"	" "	G/3/1/1/33	A	Scrapped 1976	"
F6	"	" "	"	A	Sold - 1975	"
F7	1881	Ashbury "		A	Lost - Fire 1975	"
F8	"	"	G/3333/3	A	Scrapped 1970	Roxey 4NGC6
F9	"	Brown Marshalls	3/3/3/1/1/3	A+C	Extant	"
F10	"	" "	3/33/33/3	A+C	Extant	"
F11	"	" "	"	A+C	Extant	"
F12	"	" "	"	A	Scrapped 1982	"
F13	1894	Brown Marshalls	G/3/1/1/33	A	Lost - Fire 1975	Roxey 4NGC7
F14	"	" "	"	A	Scrapped 1976	"
F15	"	" "	G/3/1/1/3/3	A+C	Extant	"
F16	"	" "	G/33/33/3	A	Scrapped 1976	Roxey 4NGC8
F17	"	" "	"	A	Scrapped 1975	"
F18	"	" "	G/3/3/3/3/3	A+C	Extant	"
F19	"	" "	G—/3/3/3	A	Scrapped 1976	Roxey 4NGC9
F20	1896	Metropolitan	"	A	Lost - Fire 1975	"
F21	"	"	G/1/1/1/1/3	A	Sold - 1975	N/A
F22	"	"	"	A	Scrapped 1976	N/A
F23	"	"	"	A+R	Extant - Runner	N/A
F24	"	"	"	A	Lost - Fire 1975	N/A
F25	"	"	3/33/33/G	A+C	Extant	Roxey 4NGC8
F26	"	"	"	A+C	Extant	"
F27	1897	Metropolitan	Luggage Van	A+E	Extant	Roxey 4NGC10
F28	"	"	"	A+E	Extant	"
F29	1905	Metropolitan	3rd Saloon	B+C	Extant	Roxey 4NGC11
F30	"	"	"	B+C	Extant	"
F31	"	"	"	B+D	Extant	"
F32	"	"	"	B+D	Extant	"
F33	"	"	G—/3/3/3	B+R	Extant - Runner	Branchlines 4.75
F34	"	"	"	B	Lost - Fire 1975	" "
F35	"	"	"	B+D	Extant (with bar)	Roxey 4NGC12
F36	"	"	"	B+S	Extant - Museum	"
F37	1899	Hurst Nelson	G/3/1/1/3/3	MNR	Ex-MNR - Sold	Roxey out soon
F38	"	"	3/3/1/1/1/3	MNR	Ex-MNR - Sold	" " "
F39	1897	Bristol & S.Wales	33/33/G	MNR	Ex-MNR - Extant	Roxey 4NGC1
F40	1907	Metropolitan	G—/3/3/3	B+R	Extant - Runner	Branchlines 4.75
F41	"	"	"	B+C	Extant	" "
F42	"	"	"	B	Lost - Fire 1975	" "
F43	1908	Metropolitan	"	B+N	Extant - Derelict	" "
F44	"	"	"	B+R	Extant - Runner	" "
F45	1913	Metropolitan	G/3/1/1/3/3	B+C	Extant	Branchlines 4.77
F46	"	"	3/3/1/1/3/G	B+C	"	" "
F47	1923	Metropolitan	3/3/3/3/3/3	B+C	"	Branchlines 4.76
F48	"	"	"	B+C	"	" "
F49	1926	Metropolitan	G—/3/3/3	B+C	"	Branchlines 4.75
F50	1925	Metropolitan	333+333	P+R	Extant - Runner	Roxey out soon
F51	1912	"	"	P	Scrapped 1968	" " "
F52	"	"	3/3/3+33/3	P	" "	" " "
F53	1919	"	1/1/1+333	P	" "	" " "
F54	1923	"	3/3/3+33/G	P+O	Being Reconstructed	" " "
F55	1912	"	333+333	P	Scrapped 1968	" " "
F56	1924	"	1/1/1+333	P	" "	" " "

Number	Date	Makers	Type	Notes	Status	4mm Kit
F57	1919	"	333+333	P+N	Extant	" " "
F58	1922	"	"	P	Scrapped 1968	" " "
F59	1920	"	3/1/1+33/3	P	" "	" " "
F60	1923	"	333+33/3	P	" "	" " "
F61	1921	"	3/3/3+333	P	" "	" " "
F62	1926	"	"	P+N	Extant	" " "
F63	1920	"	333+333	P+N	"	" " "
F64	1912	"	333+33/3	P+H	"	" " "
F65	1910	Metropolitan	33/3+33/3	P	Scrapped 1975	Roxey out soon
F66	1920	"	333+333	P+N	Extant	" " "
F67	1922	"	33/3+333	P+N	Extant	" " "
F68	1909	"	3/3/3+33/3		Sold 1975	" " "
F69	1923	"	333+333	P	Scrapped 1968	" " "
F70	1922	"	333+333	P+N	Extant	" " "
F71	1911	"	333+33/3	P+R	Extant - Runner	" " "
F72	1926	"	3/1/1+333	P	Scrapped 1968	" " "
F73	1920	"	3/3/3+3/3/3		Extant - Runner	" " "
F74	1921	"	3/3/3+333	P+N	Extant	" " "
F75	1926	"	3+3 Saloon	P+S	Extant - Museum	" " "

IMR 'N' SERIES (ex-Manx Northern Railway) CARRIAGE LIST:

Number	Date	Makers	Type	Notes	Status	4mm Kit
N40	1979	Swansea Wagon Co.	1/1/1	MNR	Sold 1975	Roxey 4NGC4
N41	"	"	3 Saloon	MNR	Extant - Mess hut	"
N42	"	"	G/3/3/3	MNR	Extant - Museum	Roxey 4NGC3
N43	"	"	"	MNR	Scrapped 1975	"
N44	"	"	G/3/33/33	MNR	Scrapped 1975	Roxey 4NGC2
N45	"	"	33/33/3/G	MNR	Sold 1975	"
N46	"	"	33/33/33	MNR	Scrapped 1975	"
N47	"	"	"	MNR	Scrapped 1975	"
N48	"	"	"	MNR	Scrapped 1972	"
N49	"	"	"	MNR	Scrapped 1975	"
N50	"	"	33/33/33	MNR	Scrapped 1975	"
N51	"	"	"	MNR	Sold 1975	"

NOTES:

A = Small 'F' bogie carriage – overall length 35' 0", height 9' 4" – wooden body and frames, the body sides were carried down to mask the sides.

B = Large 'F' bogie carriage – overall length 36' 11", height 10' 3" – wooden body and metal frames.

C = Current operational service fleet at the time of writing.

D = 'Bar Set', Saloon coaches F31, F32 and F35 were fitted with corridor conections in 1981 and F35 was also fitted with a bar and toilet.

E = 'Empress Van' full luggage/brake bogie carriage – during their lives, these useful vans have been used on all sorts of duties including moblie PW bothy and store sheds and even as an ambulance during the famous TT motorcyle races!

H = One half (C1 body) of 'Pairs' F64 remains resting on a unidentified 'M' wagon in Douglas carriage shed.

N = Not in use – all stored in Douglas carriage shed waiting restoration to running order at the time of writing.

MNR = Ex-Manx Northern Railway stock Nos. 1 to 17 absorbed in 1905 by the IMR. Both the ex-MNR bogie – and six-wheeled coaches were allocated IMR. 'F' series Nos. F37 to F51, but with the arrival of further new bogie – coaches, the MNR six-wheeled Cleminson coaches were renumbered N40 to N51.

O = Being completely reconstucted in Douglas workshops and. as no plans exsist, a picture in a railway book is being used as the main reference source!

P = 'Pairs'. These were the original 'A', 'B', 'C' and 'D' series four-wheeled coach bodies mounted in pairs on a brand new bogie chassis supplied by Metropolitan. The re-building took place between 1909 and 1926, while the redundant four-wheeled chassis were utlisied to construct new goods and cattle wagons!

R = 'Runners'. A number of carriages have had their bodies removed and scrapped. The remaining chassis is used for various uses including to carry containers in 1967, for P.W. Work as rail carriers and one was converted into a ballast hopper wagon. In 1968 several of the runners were numbered R1 to R11.

S = F75 – known as the 'Governor's Saloon' and F36 – 'Royal Saloon' are on display in Port Erin Railway Museum. F36 was last used in September 1993 to carry the Duchess of Kent.

IMR F28 'Empress van' at Douglas station on March 27 1994.

reinstated – N41, 43, 48 & 51 remained electrically lit. The IMR. marshalled the carriages into rakes with a dynamo-fitted coach and a number of slave coaches, normally six short four-wheelers or three 'F' series bogie-carriages. As traffic dropped off in later years the coach lighting fell into disuse, all late-night special trains now have their lights powered by temporary car batteries.

Interiors:
Originally varnished wooden panels with head-high upholstery, curtains and hat racks in 1st class compartments and bare wood seats and not a lot else in the 3rd class. All compartments are now upholstered throughout the current service fleet.

Steam heating:
To improve comfort during the winter months, the IMR. fitted steam heating to 26 of its 'F' series carriages in 1937 – prior to this the only heating was provided by second-hand footwarmers. As with the lighting, this also fell into disuse in later years.

Vacuum braking:
The need for full continuous braking on all passenger train was high-lighted in 1925, with a fatal crash involving a runway train at Douglas station. Prior to this, all braking was done by means of handbrakes only. Only the carriages with guards compartments had handbrakes, all the other coaches had no brakes at all – most trains having a guard at the rear and a brakesman in the front coach. The resulting inquiry recommended the fitting of continuous brakes. However, the fitting of such brakes was tackled at a very slow pace and by 1950, the task still remained incomplete – eight 'F' series coaches were never equipped including the two 'Empress Vans'. The problem with the steam heating and the vacuum brakes was because IMR engines, boilers could not produce enough steam to sucessfully operate both, either or either systems. However in recent years, with the advent of larger boilers and shorter trains, full continuous braking has been fitted to all the current operational rolling stock, including one of the 'Empress vans' F28.

IMR & MNR CARRIAGES IN MODEL FORM

Today, thanks to the efforts of Messrs Branchline and Roxey Mouldings, it is possible to construct nearly every IMR and MNR coach in 4mm scale. As you can see on the main carriage list there are very few gaps remaining in the list of 4mm scale models. Prior to this - recent prolific introduction of new carriage kits over the last three years, there was only the old Gem white metal small 'F' bogie coach and four-wheeler carriage available in kit form. While, many IMR modellers had to resort to scratch-building or a major conversion job on a proprietery item to build up a decent fleet of IMR carriages. Meanwhile, with the 4mm IMR carriages almost complete, Roxey Mouldings has turned to the hugely popular 7mm scale. Its first kit in this scale was the ex-MNR 'Foxdale Coach' F39, introduced a couple of years ago. F39 is to be joined in 7mm scale very shortly by the ex-MNR six-wheeled 'Cleminson' carriages – N40 to N51 (MNR No.1 to No.14). If there is enough interest generated in modelling the IMR & MNR in this scale, Roxey Mouldings could well produce other items of rolling stock in the future. Also on the drawing board, but in a much larger scale of 15mm to 1', D. J. B. Engineering is planning to produce IMR carriage kits in this scale in late 1996/early 1997. However, Priory Carriages already produce a fine range of ready-to-run examples of MNR coaches including the Foxdale coach F39 and a MNR luggage/brake. IMP Models also produces semi-scale vacuum moulded kits of the early IMR four-wheeled coaches and one of the 'Pairs' bogie-carriage, all in 15mm to 1' scale.

As you can see, the Manx modeller is now well catered for in the essential passenger stock department leaving very little for the scratch-builder to do.

Chapter Five

IMR & MNR WAGONS

Unlike most railways, the Isle of Man Railway depended almost entirely on the passenger receipts during the short hectic summer season to make its profit. Goods traffic was very limited and a poor source of income. There were never any private owner wagons on the Island, only railway company vehicles. Pure goods trains on the Manx railways were a very rare sight indeed and usually consisted of cattle show specials, which was the one of the few occasions that the locomotive *Caledonia* would be steamed up and pressed into service.

Any goods traffic was normally attached to the rear of the passenger service trains. A small footnote in the timetable stated: "Trains will pick up goods wagons on the road, and will be liable to delay in consequence". Shunting of goods traffic away from the towns in the smaller country stations was often done by either hand- or fly-shunting or a combination of both! This practice was not permitted on the mainland, but the IMR was a law unto itself. A few of the IMR and MNR staff have been seriously or fatally injured while shunting over the years, but certainly no more than on a 'safer' mainland railways.

The IMR and ex-MNR wagon fleets consisted almost entirely of four-wheeled stock. It is only in the last 30 years that freight bogie-wagons have appeared on the Island's railway. These all consist of ex-carriages after the bodywork was scrapped.

The IMR wagon fleet was identified by a number prefixed by a letter. These followed on from the 'F' series bogie carriages. All ex-MNR stock had a small 'R' after the letter, i.e. G[R]10 to denote that it was former Manx Northern Railway stock. As with the locomotives and carriages, the IMR and the ex-MNR wagon fleets are very complicated and a detailed study of these wagons would fill this book.

In 4mm scale, with the locomotives and carriages very well catered for, the wagon fleet offers the most opportunity for the scratch-builder at present, with only couple of the wagon kit available in this scale. The old Gem 'H' wagon and 'G' van can still be picked up second-hand, while Branchlines has produced both the 'H' and 'M' wagons in 4mm scale recently. A 'K' series cattle wagon and a 'G' van are due to follow from Branchlines. These new kits can form the basis of many conversions to add lots of variety to your roster of IMR wagons. Roxey Mouldings is planning to bring out four-wheeled coaches in 4mm scale. The chassis of these kits could be used to scratch-build the many variations of 'mongrel' vans built by the IMR on these redundant chassis. There are no IMR or ex-MNR wagons available in 7mm scale just yet – but I am sure it will not be long. In 15mm to 1', IMP models has produced a semi-scale kit of an IMR 'G' van and DJB Engineering has several items of both IMR and MNR rolling stock in the pipeline for release in1996-97.

'G' van G16, built in 1915 by IMR, on chassis C1, at Peel station in August 1955. LES DARBYSHIRE

THREE TYPES OF IMR WAGONS

A) The original wagons purchased from various manufactures such as Metropolitan, Brown Marshall and Ashbury for the IMR and Swansea and Hurst, Nelson for the MNR.

B) Those built 'in-house' by IMR and MNR in their own workshops using bought-in metalwork and wheelsets. Their outlines were very similar to the above original designs, but with slight minor differences/improvements.

C) 'Mongrel' wagons. To reduce marshalling time, the bodies of early four-wheeled coaches were in the early 1900s, mounted in pairs on new bogie chassis and give 'F' series numbers F50-F75, while the redundant four-wheel chassis of the coaches was used to construct various covered and open goods, fish wagons and cattle vans. All were built to similar design of the original wagons. They could be quickly identified from the other wagons as they were all 2' longer than the original types and had coach-type axleguards and springing.

IMR breakdown crane No.2, 8-Ton capacity, built by Richard C. Gibbins & Co. in 1895. Seen here at Douglas in 1985, it is now on display at Union Mills station on the 'Railway Heritage' trail.

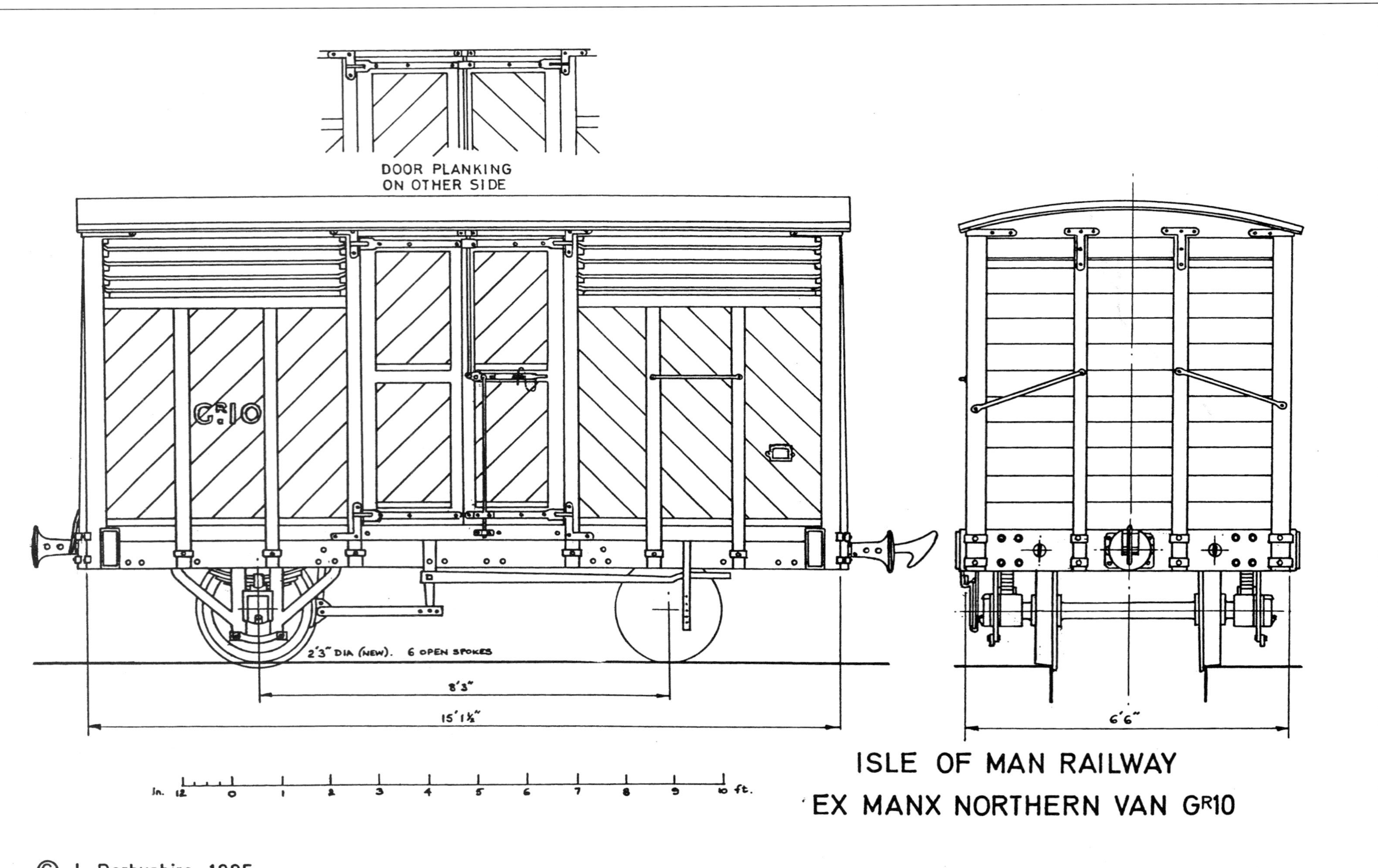
DOOR PLANKING
ON OTHER SIDE
GR10
2'3" DIA (NEW). 6 OPEN SPOKES
8'3"
15'1½"
6'6"
In. 12
0
1
2
3
4
5
6
7
8
9
10 ft.
ISLE OF MAN RAILWAY
EX MANX NORTHERN VAN GR10
© L Darbyshire 1995

IMR + EX-MNR WAGON LIST

Series G = G1-G19 - four-wheeled, covered goods van.

Number:			Builder:	Year:	Notes:	Remarks:
G1	-	G4	Metropolitan	1873	A	
G5	-	G6	Ashbury	1877	A	
G7			IMR	1897	B	Rebuilt from smashed 1873 E1 remains.
G8	-	G9	Metropolitan	1899	A	
G^{R}10	-	G^{R}13	Swansea	1879	A	Ex-MNR Nos.13-16, all had new bodies in 1896.
G^{R}14			Swansea / MNR	1896	B	Rebuilt from MNR No.15 passenger brake van.
G15	-	G19	IMR	1915-21	C	Built on ex-four-wheeled coach chassis.

Series H = H1 - H46 - four-wheeled, three-plank drop-door open wagon.

Nos:			Builder:	Year:	Notes:	Remarks:
H1	-	H20	Metropolitan	1873	A	Fitted with rubber block spring.
H21	-	H26	Ashbury	1877	A	Fitted with coil springs.
H^{R}27	-	H^{R}38	Swansea Wagon	1879	A	Fitted with leaf springs.
H^{R}39			Ashbury	1880	A	Shorter wheel-based oddity.
H^{R}40	-	H^{R}45	Hurst Nelson	1900	A	Fitted with leaf springs.

Series K = K1-K26 - four-wheeled cattle van.

Nos:			Builder:	Year:	Notes:	Remarks:
K1	-	K2	Metropolitan	1873	A	
K3	-	K4	Ashbury	1877	A	
K5	-	K6	Metropolitan	1899	A	
K^{R}7	-	K^{R}9	Swansea	1879	A	Ex-MNR Nos. 17 - 19
K10	-	K12	IMR	1908	B	Built on ex-M43 - M45 chassis.
K13	-	K14	IMR	1912	C	Built on ex-four-wheeled coach chassis.
K15	-	K16	IMR	1912	C	Built on ex-four-wheeled coach chassis 'G' body.
K17	-	K18	IMR	1914	C	Built on ex-four-wheeled coach chassis 'G' body.
K19	-	K20	IMR	1920	C	Built on ex-four-wheeled coach chassis 'G' body.
K21	-	K22	IMR	1921	C	Built on ex-four-wheeled coach chassis.
K23	-	K26	IMR	1923	C	Built on ex-four-wheeled coach chassis.

Series L = L1-L4 - four-wheeled, bolster wagon.

Nos:			Builder:	Year:	Notes:	Remarks:
L1	-	L4	Metropolitan	1874	A	Rubber blocks, later replaced by laminate springs.
L5	-	L6	IMR	1910	C	Built on ex-four wheeled coach chassis.

Series M = M1-M78 - four-wheeled, two-plank drop-side open wagon.

Nos:			Builder:	Year:	Notes:	Remarks:
M1	-	M4	Ashbury	1877	A	Early type.
M5	-	M7	Ashbury	1884	A	Early type.
M8	-	M19	Ashbury	1888	A	Early type.
M20	-	M27	Metropolitan	1899	A	Early type.
M^{R}28	-	M^{R}35	Swansea Wagon	1884	A	Early type, ex-MNR 22 - 29.
M^{R}36	-	M^{R}37	Manx Northern Ry	1894	A	Early type, ex-MNR 30 & 31.
M^{R}38	-	M^{R}39	Manx Northern Ry	1896	A	Early type, ex-MNR 32 & 33.
M^{R}40	-	M^{R}42	Manx Northern Ry	1898	A	Early type, ex-MNR 34 - 36.
M43	-	M54	Metropolitan	1911	A	Later improved type.
M55	-	M60	Metropolitan	1924	A	Later improved type.
M61	-	M72	Metropolitan	1925	A	Later improved type.
M73	-	M78	Metropolitan	1926	A	Later improved type.

Series R = R1-R11 - bogie container wagons, ex-F series chassis.

Nos:			Builder:	Year:	Remarks:
R1	-	R11	IMR	1967-68	Ex-bogie chassis of F51, F52, F53, F55, F56 F58, F59, F60, F61, F69, F72.

Four-wheeled Fish Wagons - not taken into IMR letter series.

Nos:			Builder:	Year:	Notes:	Remarks:
1	-	2	IMR	1909	C	Built on ex-four-wheeled coach chassis.
3			IMR	1910	C	Built on ex-four-wheeled coach chassis.
4	-	5	IMR	1914	C	Built on ex-four-wheeled coach chassis.

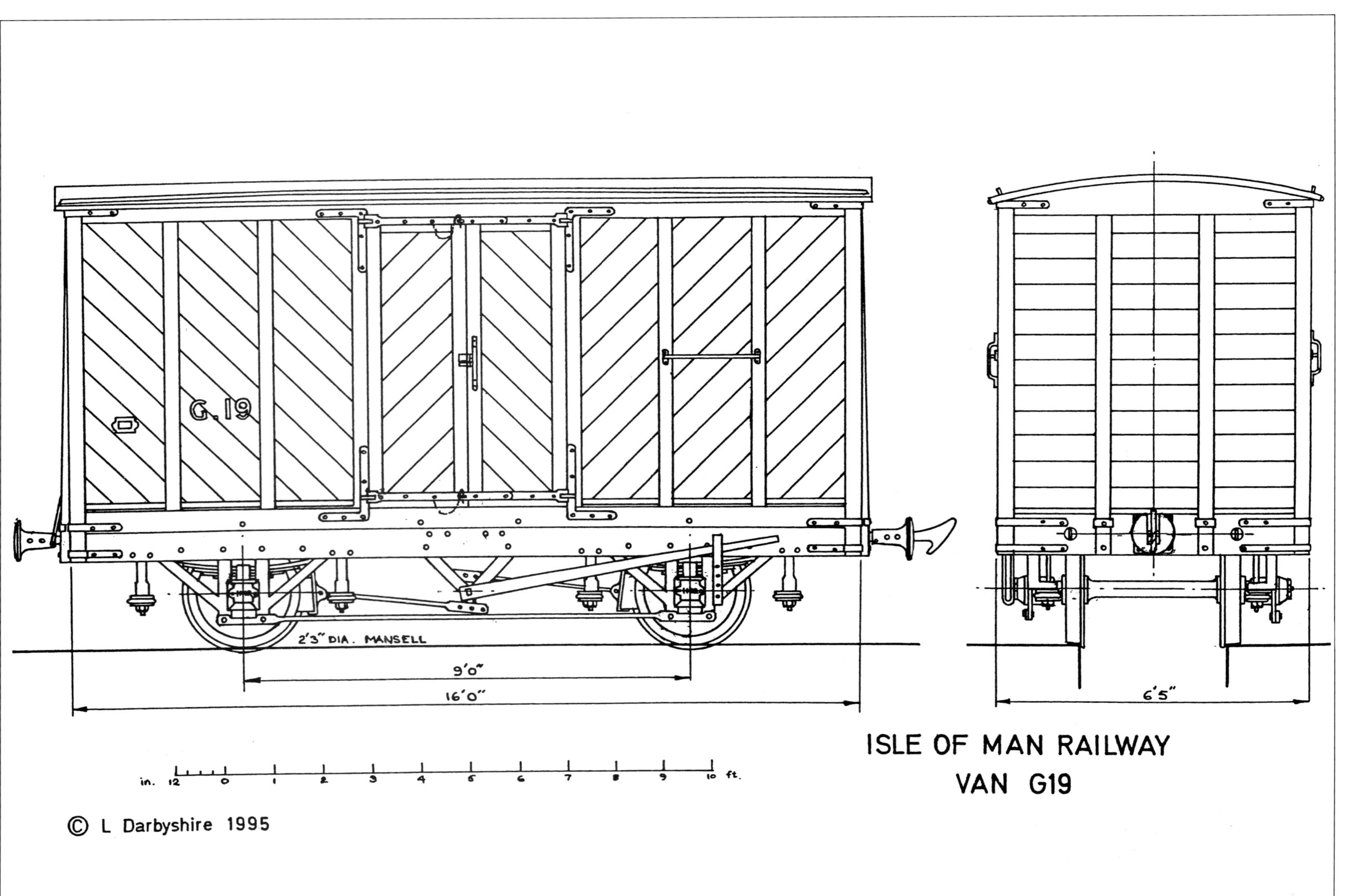

G.19
2'3" DIA. MANSELL
9'0"
16'0"
6'5"
in. 12 0 1 2 3 4 5 6 7 8 9 10 ft.
ISLE OF MAN RAILWAY
VAN G19
© L Darbyshire 1995

Two-plank dropside ballast wagon M56, built by Metropolitan in 1924, seen at Douglas in July 1966. A kit is available for 4mm from Branchlines and a 15mm version is promised from DJB Engineering.
LES DARBYSHIRE

Ex-MR 'G' van G^{R}10, built by Swansea in 1879, at Douglas station in 1966. The vehicle was scrapped in 1974. 4mm and 15mm scale versions are available from Branchlines and DJB Engineering respectively.
LES DARBYSHIRE

A good selection of IMR wagons in Douglas station's goods yard in September 1969. These include a 'R' bogie flat, two mongrel 'G' vans built on ex-four-wheeled coach chassis and an early IMR 'G' van. Another van is an ex-IMR brake/luggage van, E1, now used as a goods van. One of the 'Empress Vans' F28 is also visible in the far left of the picture.
TREVOR NALL COLLECTION

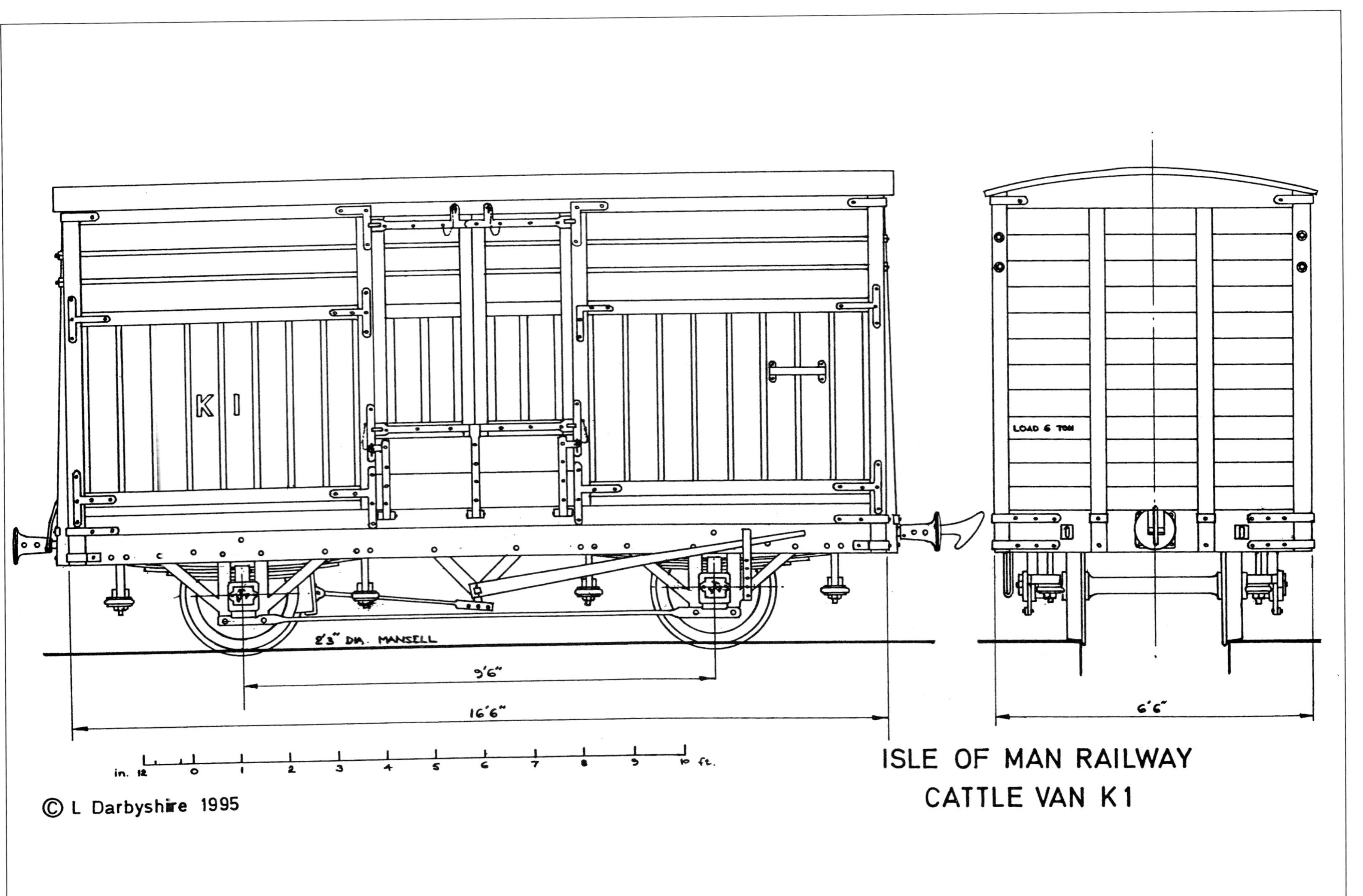
K1
LOAD 6 TON
2'3" DIA. MANSELL
9'6"
16'6"
6'6"
in. 12 0 1 2 3 4 5 6 7 8 9 10 ft.
ISLE OF MAN RAILWAY
CATTLE VAN K1
© L Darbyshire 1995

Chapter Six

STATION BUILDINGS & OTHER RAILWAY STRUCTURES

The main terminus and headquarters of the Isle of Man Railway is situated at Douglas. In its heyday, Douglas was a narrow gauge station second to none and was the envy of all other narrow gauge railways – and quite a few standard gauge railways as well! An imposing entrance leads into a large forecourt formed by the station building and main offices, which are both in red Ruabon brick. Beyond these are two very long canopied Island platforms, each capable of holding two complete trains coupled together!

The two outer faces of the two platforms were normally used for departures, while incoming trains used the inner faces, so the engines could run-round their train using the central release road in between the two platforms. A coal stage and water column were provided at the end of this release road, so engines could be rapidly re-coaled, watered and oiled as some of the turn-round times in the timetable could be very tight indeed – keeping the footplate crew on their toes, especially if they were running a bit behind with their previous train. Other facilities includes a two-road locomotive running shed capable of holding a majority of the Manx engines. Also, due to the isolation of the Island, Douglas has a fully equipped self-sufficient workshop so complete overhauls of the locomotive fleet can be performed 'in-house', a large four-road carriage shed and goods sheds and one of the Islands two signal boxes (the other was at St John's station).

Turning the clocks back almost 125 years to the opening of the line in 1873, the original Douglas station was constructed on the same site as the present one. The whole area was reclaimed marshy ground made by the two rivers Dhoo and Glas – hence the name of the town.

Due to financial shortages, the old Douglas station building was a simple wooden affair, being a slightly larger version of the out of town stations. Other buildings such as the engine, carriage and goods sheds were of a similar basic design. The original track layout at Douglas was woefully inadequate. The IMR proved to be far more popular than first anticipated in the original surveys. A programme of improvements to enlarge Douglas to the ultimate narrow gauge station took place between 1875 and1929. Today, with only the Port Erin line surviving, Douglas's grand station has been rationalised to a half of its former self. Gone is one of the platforms and both canopies and all the goods facilities to make way for a car park. All the semaphore signal have long disappeared and the unique signalbox and large carriage shed have both seen better days. Robbed of some of its atmosphere, Douglas, is still a very impressive station. However, plans are in-hand to replace some of the signals, re-locate the signalbox closer to the remaining trackwork and building a new carriage shed, which will be of a design inkeeping with the rest of the railway. Hopefully, these improvements will go some way to regain some of the flavour of the good old days. Out of town, the stations are very simple, usually consisting of a long passing loop, a couple of sidings and a basic station building and goods facilities. Building styles vary from line to line, the IMR stations and Crossing Keeper's shelters on both the Peel and Port Erin lines were original simple wooden structures as funds were limited.

The good old days when Douglas was still an impressive station with the two long island platforms, full signalling and large goods shed on the right. Viewed from Douglas signal box in July 1971.
TREVOR NALL

Douglas signalbox, now out of use and not really part of the station any more. There are plans to move the box to the other side of the station so that it can be used with the rest of the remaining railway.

As the IMR prospered, some stations received stone or brick station and goods buildings, while many such as Santon retain their original wooden structures even today. The two other IMR termini of Peel and Port Erin were soon altered. Both received new station buildings and goods and engine sheds to cope with the ever increasing traffic demands. Port St Mary Station on the south line has for some reason a huge brick station building, which is totally out of keeping with the size of place. In fact, the station is over half a mile from the village itself – Port St Mary station is actually nearer to Port Erin! In recent years, a couple of the stations on the remaining south line have either been refurbished or replaced. Due to re-development in and around Ballasalla Station, a brand new brick station building and full-height platform have been built on the opposite side of the tracks. This new 'Thomas the Tank Engine' style building replaces a classic IMR wooden one. Castletown station building has not been replaced, but has had a major facelift in the last couple of years. Colby Station has lost its original 1874 station building, which has been replaced by the small wooden hut from former Kirk Braddan halt on the old Peel line.

On the ex-MNR Ramsey to St John's line, all the station buildings and Crossing Keeper's house are all built out of red sandstone to a similar design and although the line has been closed for many years, all still survive mainly as private houses, although Kirk Michael station building and goods shed are now the village fire station. While Ramsey station – the headquarters of the former Manx Northern Railway – has all but disappeared to make way for a large bakery plant. Various drawings of some of these buildings have appeared in different publications over the years. Apart from Douglas Station with its large frontage, most building are on the whole quite simple affairs, which a modeller could construct without too much difficulty. Scratch-building is the only option at present. However somebody may bring out kits of IMR and MNR railway buildings in the future.

Drawings only: Foxdale station building – 'Manx Northern Railway' published by Hillside Publishing
Original Douglas station building – 'Isle of Man Railway Vol. 1'
Original Peel station building – 'Isle of Man Railway Vol. 1'
Crosby and St John's Station Building – 'Isle of Man Railway Vol. 1'
Original Ballasalla, Colby, Port St Mary and current Santon station building – 'Isle of Man Railway Vol. 1'
All published by Oakwood Press.

Drawings and model construction articles:
Crossing Keeper's hut – 'Model Railway News', November 1969
Colby station building – 'Model Railway News', May 1970
St John's signalbox – 'Model Railway News', August 1970
Union Mills station building – 'Railway Modeller', December 1963
St John's station building – 'Railway Modeller', November 1967
Kirk Michael goods shed – 'Railway Modeller', November 1967
St John's ground frame hut – 'Railway Modeller', October 1968

The 'new' Port Erin station building of 1905 on the site of the old one. Photographed in May 1995.

A general view of Douglas station in 1995. The station is now half its former self with only the Ramsey/Peel platform remaining. The station still has a certain charm and remains the envy of many other railways.

The business end of Douglas station in 1995, with the station building on the left and main offices on the right. The 'rocket'-shaped corner of the main offices was the General Manager's office, where he could keep an eye on the station. The dark wooden fence marks approximately where the Port Erin platform was situated. Both island platforms had canopies. The palm trees are a feature of the Island's stations.

Douglas station building of 1891, seen here in 1995. The offices and refreshment rooms were built in 1887.

Former MNR St Germains station building of 1879. Built in red sandstone, it has been a private house for many years. Snow is a rare scenic effect on the island!

Santon station on the IMR south line retains its original 1874 building with its very simple layout. The old 'buried' type of track is seen on the left-hand side of the loop.

Ex-MNR station at Kirk Michael on the old Ramsey line. The station and goods shed are now a fire station. Remnants of track survive in the road surface.

The original Crossing Keeper's hut of 1874, just south of Castletown station. The gatepost has been produced from three sections of rail.

Water tank at Ballasalla. Down trains used to take on water here, while Up trains would be watered at Castletown. The name on the tank is a recent idea.

The rear of Castletown station on the south line, built in 1874. Refurbished in 1994, this was a one-off IMR building.

Chapter Seven

THE MANX PERMANENT WAY

The permanent way on the Manx railway systems is rather unusual, as it has been normal practice to ballast nearly up to rail height, all but covering the sleepers. On the mainland this practice was frowned upon by the British Board of Trade inspectors who said it rotted the foot of the rail and wore away the edge of the wheel flanges, but the Isle of Man is not under their direct jurisdiction. In fact, as most of the island stations had no platforms, all the trackwork in the station areas was completely buried, appearing as two groves in the mud – three foot apart!

However, in recent years, the Isle of Man Railway has adopted the more normal practice of having the rails and sleepers exposed, though in quite a few places the buried trackwork still persists. The actual trackwork is laid in typical narrow gauge fashion with 'wiggley' straight sections and 50p curves – however bad your track-laying is it will never be as imperfect as the prototype. As the IMR used to operate a rather snappy and intensive service, the little Manx narrow gauge engines would often reach speeds of 40mph plus on the lightly-laid track. Replacement of broken springs on the engines and rolling stock was an all too frequent job for the 'Workshop Boys' in Douglas's workshops.

The state of the track and the ride was notorious in the old days. This much-told famous IMR tale sums up the general feeling on the subject: A local man from Castletown arrived at Douglas station one day with a large amount of luggage. The Douglas Station Master enquired if he was "going across", to which the reply was he had a much longer voyage, as he was emigrating to New Zealand and added; but thank God that the worst of the journey is over!

As with the ballasting, the trackwork is currently being up-graded to give the passengers a more comfortable ride, instead of them being part of the suspension. Although to many, rocking and rolling your way down to Port Erin is part of the charm of the Isle of Man Railway!

The flat-bottomed rail was originally 45lb weight, but this was found to be too light. The rail section was increased over the years to 56lb. Today, a majority of the rail is either 60lb, or more recently, 75lb ex-Channel Tunnel stock in 30' lengths. The rail is, in the main, dog-spiked to sleepers. However, the whole of the Foxdale branch and areas such as tricky curves and overbridges has flat-bottomed rail mounted in chairs. The first sleepers on the IMR were half round section. Replacement sleepers were the more normal full type being 7' long, 9" wide and 5"deep. Recent supplies have been short-sawn standard gauge sleepers.

The practice of burying the trackwork is very useful, as it can be used to hide a multitude of modelling sins! Original Manx layouts

No.10 *G. H. Wood* of 1905 backs down to take out 2.10pm service train to Port Erin on May3 1995. Note the old starter signal post with arms removed and colour lights installed – it is planned to replace the signal arms in the near future.

A Stevens & Son signal lever survives in use.

tended to use TT (12mm gauge) track with the sleepers spread out more to represent narrow gauge trackage. The Isle of Man Model Railway Group's 'Castletown' layout first featured hand-built track using a scale code 60 rail. This, like the prototype's first rail, was found to be too light, which caused countless running problems. Following the prototype, we ripped up the original and relaid the whole layout with commercial track and pointwork to a larger code 80 section.

Commercial track and pointwork is available from Bemo/Shinohra or 3mm Scale Model Railways, who can supply not only flexi-track and a very large range of points, but they can also build pointwork to the customer's own requirements. As for the ballast, or should I say mud, this is best represented with fine granite ballast for N gauge, oversprayed with various greys and browns to tone it down. Finally, do not forget to add the grass and weeds that grow over the track. All the Manx coaches have double footboards for use at the many out-of-town stations without platforms and not – as first appears – to cut a swathe through the vegetation that encroaches onto the track from the hedgerows!

SIGNALLING

Both the Isle of Man Railway and Manx Northern Railway were comprehensively signalled throughout to protect their system's single-line working with passing loops and the many road crossings which punctuated the routes. The signalling equipment came from different suppliers. The original IMR signals were the slotted-post type with revolving coloured lamps supplied by A. Linley & Co. of Birmingham. However, the MNR had more the conventional type built locally using parts supplied by Messrs. Stevens & Sons. Later

IMR signal (loop) home windlass at Port Soderick station used to control entry of Up trains into the passing loop, photographed in 1996.

types of IMR signals were also supplied by Messrs. Stevens & Sons. For identification purposes, the IMR had plain ends to its signal arms, while the MNR had more distinctive and ornate fish-tail ends.

As Douglas, the main terminus, was handling up to 100 trains daily during the heyday of the IMR in 1930s, full signalling was therefore necessary. This was all controlled from one of the Island's two signalboxes. The 36-lever signalbox and all the signalling equipment controlling movement within Douglas station was supplied by Dutton & Co. of Worcester in 1892. Engines would attract the attention of the signalman by means of a special engine whistle code. Each movement within the station had a engine whistle code. For example, leaving the engine shed for No. 1 line – four short whistles or leaving No. 7 line – one crow between two short whistles and leaving line No. 6 for 'right-away' for Port Erin – two long and one short whistle.

The other slightly smaller signalbox was at St John's Station, which was the Crewe Junction of the Isle of Man. The lines from Ramsey, Peel, Foxdale and of course Douglas all converged at St John's Station. St John's joint station was also in the old days the 'frontier' post were the IMR meet the MNR. The Down trains to Peel and Ramsey would very often be combined, being either double-headed or banked. These would split at St John's, which would involve some complicated traffic movements with trains for Ramsey, Peel and the return working to Douglas all being in the station at once.

IMR original slotted-post signal just south of Castletown station, now out of use. A lamp was fitted to the top of the black shaft which revolved in conjunction with the signal arm movement.

Chapter Eight

LIVERIES: THE QUESTION OF COLOUR

No. 5 *Mona* of 1874 taking on water at Peel station in August 1964. *Mona* was fresh out of the paint shop in a smart and bright Indian red livery. Notice that the lining has been missed off the axlebox cover and the IMR practice of painting the coupling rods white. BARRY C. LANE

The question of colour has caused countless arguments for years. The human eye is a fickle instrument, with very few of us having near perfect eye-sight. In fact a large percentage suffer from some sort of colour blindness in varying degrees. For example, if you show ten people a colour sample, they will see it in ten different shades and at least one will see it as a completely different colour! Sunlight has a habit of play tricks on eyes as it reflects of the varnished paintwork changing the hues. During the day, the natural light will vary depending on the time of day and type of weather from full sunlight on a clear day to the filtered blue/grey light of a cloudy dull day. While what we do remember of a colour, the exact shade will over the years fade with time as our memory looses its edge.

However, they say the camera never lies, but there can be many variations of colour depending on the age of film, how it is stored and the developing and printing process. While some certain types of early colour slide film are not colour-fast and will change over the years. Kodachrome films of the 1950/60s were also known for rendering a blue tone to colours – so in the end, what is really the exact colour of your favourite Manx engine?

The only true way to get the correct colour is to take a sample of the paint layers of the prototype. However, even this will not be totally accurate as the old oil-based pigment paint goes darker with age, while modern paints get lighter!

There are many good colour photographs of the Manx locomotives taken in the 1950/60s. However, each engine appears to be a different shade of Indian red. This has been caused by the old fashioned hand-mixed pigment and oil type paint going darker with age due to the heat and the variation in the colour was due to the fact that the engines were painted over a number of years and were very often patch painted or just re-varnished as required. The paint itself was very often mixed by different railway staff working in the paint shop. Cleaned with an oily rag, which would produce a nice shine, but it would also darken the paintwork – hence no two engines looked exactly the same. While we are touching on the subject of locomotive cleaning, have you ever wondered how they keep the brass dome and copper pipework on the Manx 'Peacocks' absolutely gleaming? Simple, a good rub ever so often with 'HP Sauce' does the trick – honest!

So there are many variations of the Isle of Man railway's Indian red colour during that period. It is all down to a matter of personal taste. I say – if it looks right, then it is right.

The reverse process is happening with current liveries of the locomotives. The modern non lead-based, synthetic paints fade quite quickly, so each year the engines look a shade lighter! One person's perception of a colour that they use on their models can look totally and completely wrong to another. I have included with these livery notes some references to model paints that I personally think look right. Various members of the Isle of Man Model Railway Group and I have spent many hours studying hundreds of old photographs and taking samples from the real thing. The colours we have used on our models have been ever so slightly compromised to allow us to use commercially available paints. This, hopefully, maintains some continuity between various members' rolling stock running on the club layout. However, your perception and thoughts on the Manx liveries may be totally different to ours.

IMR LOCOMOTIVE LIVERIES:

1873 onwards:
Brunswick green (Humbrol No. 3) – with black lining, a vermilion line inside and a white line outside. The driving wheels were also painted Brunswick green.

By 1939:
Still Brunswick green (Humbrol No. 3), but the running gear was now painted black and the lining was a red or orange line either side of a black line. Sometime after World War One, the engines started to be out-shopped with their numbers painted on the rear of the coal bunker. There are very few colour photographs of the IMR engines in this green livery. Interestingly, there seems to be a number of variations in the exact shade of green over the years. Test rubbings on several engines by the railway staff to find the correct shade has concluded there were at least four and maybe even more.

A colour photograph taken just after the war, shows one of the

Empress van F27 of 1897, built by Metropolitan, at St John's in August 1964. F27 has a very faded blood red livery, which is now a salmon pink. Behind F27 is a mongrel 'G' van built on an ex-four-wheeled coach chassis. The building on the left is part of St John's station building. BARRY C. LANE

Small 'F' F13 of 1894, built by Brown Marshall, at Douglas station in August 1964. F13, and her numerous sister carriages in the background, are in the late 1940s' 'blood and custard' livery – very similar to the BR coach livery of that time. Note the practice of painting a dummy black solebar on the coach side, so that it matched the large 'F' stock with exposed solebars! Also visible in this shot are the cables for the electric lights that ran along the carriage roofs. BARRY C. LANE

Ex-County Donegal Railcars No. 20 with No. 19 behind at Douglas station in 1995. Their current condition is poor, though due to their popularity there are plans to refurbish the two units.

locomotives, No. 16 *Mannin* at Douglas in 1949 – painted in a unlined Brunswick green – possibly a wartime austerity livery, (and a good one for modeller who hates lining!).

1945 to 1965:
Indian red (Humbrol No.100) – with black lining and with a yellow line either side. The driving wheels were also painted Indian red and the coupling rods were painted either red or white. Re-painting of the locomotives into this livery was done over a number of years, by 1947 only engines No. 3, 4, 6. had been painted Indian red. Although, most of the engines still remaining in service had been re-painted Indian red by 1950. Check photographs of individual engines for the shade of this darker aged Indian red (try Humbrol No. 180).

1967 to 1980:
Apple green – with black lining and with a white line either side – similar to the London & North Eastern Railway livery. The story behind this livery was that it was copied off a 4mm to 1' GEM model of the Isle of Man Railway 2-4-0 locomotive, built by the Rev. Teddy Boston! The new management liked it so much, they decided to repaint all the locomotives in this livery for the reopening of the railway in 1967.

Above and below: Current lettering on carriage F15.

Current liveries – 1980 onwards: The current service locomotive fleet are painted in a representation of the railway's historical liveries, all except No.12 *Hutchinson,* which was painted in a un-prototypical livery of dark blue in 1980, when it received a new larger boiler and sidetanks, also a square shaped cab. Both the livery and cab were, at the time, very un-popular with enthusiasts. Yet, after 16 years of being in this condition, both the livery and cab are now historic. However, the railway are planning to replace both at its next major boiler overhaul.

Current liveries for the Manx modern image modeller (as 31/05/96):

No. 1 1967 IMR green – on display in Port Erin museum.
No. 2 Dismantled and scrapped - 1951.
No. 3 1873 IMR green – on display in Science and Industry museum Manchester.
No. 4 Manx Northern Railway Tuscan red – withdrawn from service.
No. 5 1967 IMR green – stored in Douglas carriage shed.
No. 6 1945 IMR Indian red – after cosmetic restoration in 1994.
No. 7 Dismantled 1945 – chassis on display at Castletown station.
No. 8 1967 IMR green, dismantled. Being rebuilt?
No. 9 Painted in 1993 – un-prototypical light green, stored in Port Erin engine shed.
No. 10 1873 IMR green – currently in service.
No. 11 1945 IMR Indian Red – currently in service.
No. 12 Painted in 1980 in un-prototypical dark blue – currently in service.
No. 13 Was painted IMR 1873 green up to 1993 – dismantled, stored in Douglas engine shed.
No. 14 Manx Northern Railway Tuscan red – privately owned on the Island.
No. 15 Manx Northern Railway Tuscan red – currently in service.
No. 16 1945 IMR Indian red – unlined, on display in Port Erin museum.
No. 17 1873 IMR dark green with red chassis frames – currently in service.

IMR RAILCARS AND DIESEL LOCOMOTIVE LIVERIES:

The two ex-County Donegal Railways railcars arrived on the Island sporting the former CDR livery of red, with a cream waist band separated by a thin black line. A cream V-stripe carried on the cab fronts was of two differing styles. They remained in this livery until 1979, when they were re-painted coach red with off-white uppers with no lining and the distinctive V-stripes.

A 5.5mm scale scratch-built model by Brian Caton of IMR large 'F' bogie coach F46 in 1950's livery.

Brian Caton's 5.5mm scratch-built model of Manx Northern Railway 4-wheel luggage/brake No. 20 in MNR livery.

No. 1 *Sutherland* of 1873 at Peel station, August 1964. Compared with freshly-painted *Mona*, *Sutherland* had been painted in the same Indian red livery many years previous. This aged Indian red has gone darker with age, a good example for fans of weathering. BARRY C. LANE

The 0-4-0 diesel locomotive was supplied from Germany in 1992 with dark green bodywork and red chassis and a large white 208 number on the cab sides. It re-appeared from the IMR paintshop in 1873-1945 IMR dark green with full white/black/white lining. The red chassis was retained. No. 17 was applied to the traditional IMR position on the cab rear, while the *Viking* nameplates were attached to the cab sides.

CARRIAGE LIVERIES:

Over the years, the IMR carriage stock has carried many different liveries and like the locomotive stock needs to be studied very careful to model it accurately. Reference to good photographs is therefore an absolute must, although it is hard to find photographs of the stock in the earlier years as most seem to be of the locomotives.

A detailed study of the IMR coach liveries would easily fill a whole book, so I will try to keep it simple. Information on the liveries of the early four-wheeled stock delivered in 1873 is slightly uncertain, one published source states that the 'A' series was painted with mid-green lower panels, with off-white uppers and yellow lining. While the 'B to D' series were chocolate brown with yellow lining. This is type of livery was typical of many carriage builders of this period including the supplier – Metropolitan.

However, recently, the end of one the 'C' series – C1 was exposed when the body was removed from its 'Pairs' chassis. This end was originally painted mid-green and not chocolate brown as stated – so who can say what the initial livery of these early IMR carriages was!

The bogie stock 'F' series carriages first started arriving on the Island in 1876, sporting a off-white upper panels and chocolate brown lower panels livery. Lettering and numbers were gold shaded blue. All the four-wheeled stock was gradually re-painted in this bogie stock livery.

From about 1917 onwards, the off-white upper panels were replaced by light brown, which was not dissimilar to the livery of Lancashire & Yorkshire Railway – a number of coaches remained in these colours until the mid-1940s. The new larger 'F' series carriages purchased between 1923-26 were supplied with white uppers and purple lake lower panels with red and yellow lining. Any of the older stock was also gradually re-painted in this livery which was the same as the London & North Western Railway colours. During World War Two, any re-painting was done in an all over dark red-brown livery.

After the war, the railway under went a total livery change, with the carriages now appearing in a overall deep red, which was soon changed to deep red lower panels with off-white window panels – this livery being very similar to the early British Railways coach livery of blood and custard. This is one of the most interesting and colourful periods of the IMR to model, as both the engines and carriages appear in pre-war, wartime and post-war liveries. Furthermore, there are many colour photographs of the late 1940s, 1950 and early 1960s in books, which is a great help to the IMR modeller.

This popular post-war livery of deep red and off-white uppers persisted right up to the late 1980s, when the current running fleet was re-painted back into the old livery of white upper panels and purple lake lower panels. The carriages look very smart when first out-shopped in this livery, but weather badly after a couple of seasons. Remaining stored and out of use carriages are still in the deep red livery.

The roofs are the most noticeable part of any model carriage due to the normal viewing height of a layout. On the IMR, coach roofs were usually painted white or very light grey. Since the early 1960s, a number of carriages have appeared with black roofs. However, regardless of the original base colour, the roof very soon became a uniform smoky grey colour.

WAGON LIVERIES:

After the complex locomotive and coach liveries, the wagons are very simple – all the IMR and later MNR wagons were painted light grey, apart from the early MNR ones that were painted dark brown (possibly creosoted for cheapness). LMS goods stock grey is very close match to Manx wagon grey, although shades do vary quite a bit. The ironwork was black or grey in later years. Lettering was white, shaded black with white load and tare painted on either one or both ends. Again, in later years as an economy measure, the shaded letters were replaced with just plain white ones. As the goods traffic started to decline, many wagons were just left to rot away. The other lucky wagons left in service were in a bad state of repair and heavily weathered.

TRANSFERS:

To complete your rolling stock, Blackhams are now producing IMR transfers in 4mm scale. The range includes IMR crests, coach numbering, lettering and guard. Wagon lettering can be made up from 3mm size letters and numbers from the Woodhead range. To go with the 15mm to 1' scale Manx engine and projected rolling stock from DJB Engineering, Tawney Graphics are to produce a full range of IMR locomotive lining and letters, crests and numbers for the carriages and wagons in this scale.

Chapter Nine

OTHER MANX RAILWAY GEMS

This section is aimed at introducing other Manx railway gems that have definite modelling potential. As well as having an extensive 3' (914mm) gauge steam railway system, the Isle of Man is home to many other vintage transport delights. In fact, on the Island there are probably more forms of historical transport per square mile than anywhere else in the British Isles.

At the other end of Douglas's two mile long promenade, there is the unique pioneering 3' (914mm) gauge Manx Electric Railway. Electric trams from another age swish their way up to the town of Ramsey in the north of Island. The scenery on route is some of the best the Island has to offer. Almost connecting the IMR and MER at either end of the promenade is the 3' (914mm) gauge Douglas Horse Trams – the only horse-drawn passenger railway in the British Isles.

While a ten minute trip on the MER to Groudle and a short walk along a picturesque Manx glen and you will be on the doorstep of the 2' (600mm) gauged Groudle Glen Railway. This recently restored miniature steam railway only runs 3/4 mile out on to the coast and along headland – from the middle of nowhere to the edge of it, but who cares, it's an absolute delight.

Another 20 minutes on the MER will take you to Laxey station, the interchange for the 3' 6" (1008mm) gauge Snaefell Mountain Railway – the first in the British Isles. Climb 2,036' to the summit of Snaefell, which just happens to be the highest point on the Island. The 4 3/4 mile slog up the 1 in 12 gradient is by Victorian electric trams, built while electric traction was still in its infancy.

Besides these remaining Manx railway systems, there are the many closed lines to explore. The whole Island was once literally covered by railways of one form or another in the good old days. As well as the railways there are vintage buses and other road vehicles, shipping and aircraft – an Island that is almost sinking with every type of transport delight and full of modelling potential!

Groudle Glen Railway's *Sea Lion* stands at Sea Lion Rocks station in 1993.

A former Upper Douglas cable car running on the 3' gauge Douglas horse tram line. No. 72/73 is the last surviving example from this former system, being built out of the best bits of two trams. The cable tram is now battery-powered, driven by a motor from an old forklift truck. The car made its test run in 1995.

3' gauge Manx Electric Railway toastrack tramcars Nos. 16 and 33 at Douglas terminus at Derby Castle. It is also the terminus for the 3' gauge Douglas Horse Tramway which runs along the two mile long Douglas promenade.

On the 3' gauge Douglas Horse Tramway, Nigel trots along the two-mile long promenade with the vintage double-deck tram.

3' 6" gauge Snaefell Mountain Railway tramcar No.5 of 1895, climbs the last few yards to the summit of Snaefell Mountain, which at 2,036' is the highest on the island.

The 2' gauge Groudle Glen Railway just north of Douglas. 2-4-0T *Sea Lion* of 1896, winds its train along the rugged coastline towards Sea Lion Rocks station.

Tramcar No.1 of 1895 of the 3' 6" gauge Snaefell Mountain Railway, crosses the mountain road which is also part of the TT motorcycle course, on the way to the summit of Snaefell Mountain.

Chapter Ten

STATION LAYOUTS

The next problem facing the IMR modeller after constructing some rolling stock is which one of the many picturesque Manx stations is a suitable prototype for a layout. First thoughts of many modellers is to tackle Douglas station, the main terminus and headquarters of the Isle of Man Railway. The whole station is full of atmosphere with its two long island platforms and simultaneous departures of trains to Port Erin and Peel & Ramsey. It would make an attractive, but very large layout requiring vast amounts of engines and rolling stock to operate it properly. Douglas is for the more experienced as a long-term project or a potential club layout. The same could be said for St John's – the Island's nearest equivalent to Crewe – which would also require vast amounts of stock and space. Even if selective compression was used, both stations would still be major projects.

Most of all the other out-of-town stations on both the IMR and the former MNR lines are very simple, usually consisting of a very long dead-straight passing loop and couple of sidings. These are ideal subjects as an introduction to Manx modelling for both beginner and experienced modeller. They are also suitable for the modeller is who is very short of space – a very common problem due to the design of modern housing. These simple station layouts could be made modular and joined with other modellers' IMR layouts to form a whole section of line. Obviously, a bit of modellers' license needs to be applied to the passing loop. These are capable of holding a 20-coach train with ease, even at the quiet, little-used stations. Using selective compression again, the passing loop can be effectively shortened to hold ten carriages without compromising the overall effect of the station with a long passing loop.

A couple of suitable out-of-town stations which would make very good layout subjects are the IMR Santon station on the south line and the former MNR St Germain's station on the Ramsey line. I have also included a 'freelance' Manx layout that incorporates many of the best features of the prototype stations.

BALLAVAGUE STATION

With a little bit of imagineering, a tailor-made Manx layout can be dreamt up quite easily. Ballavague station is a freelance station incorporating the best features of all the Manx stations.

Its location is roughly half-way between Foxdale village and Ballasalla on the proposed, but never built line that would have completed the Douglas, St John's and Ballasalla triangle. Had this extension line from Foxdale to Ballasalla been built, the whole story of the IMR could have been completely different and very interesting. The traffic through Ballavague would have been much different compared with other lines, with more cattle and mineral traffic and short local trains, along with through Ramsey to Port Erin expresses, which would have eliminated changing trains at Douglas Station.

The two cattle markets at St John's and Ballasalla would have been linked, while in the early days, there would have been some very interesting movements and interchanging of IMR and MNR rolling stock. All heavy trains passing over this route, which has grades of up to 1 in 49 on the Foxdale–St John's section, would have required banking. An engine would be out-stationed at Ballavague; its prime purpose would be to assist trains through this section. The banker would join or drop off its charge at Ballavague. Some light engine movements would also add a bit of extra operational interest to the layout.

As this area of the Island was remote with only a few farmsteads, the railway just built Ballavague roughly half-way between Foxdale and Ballasalla. The station was simply a passing place between two main stations on this single line. In time, a small community developed around Ballavague station growing into a small village. The station was situated between two roads with overbridges, which makes natural breaks at either end of the layout.

Ballavague's station building is the same as the ones at Foxdale and St John's, while the engine shed is an IMR stone replacement for the original wooden version, blown down one very stormy night! The IMR also built a replacement brick goods shed at the same time, before the original wooden one also disappeared one dark windy night!

Signalling was the same as other Manx stations, with a pair of home signals controlling entry into the passing loop.

Imagineering or freelance modelling is a very useful branch of our hobby allowing modellers to create something a little different and quite possibly more interesting than a prototype station.

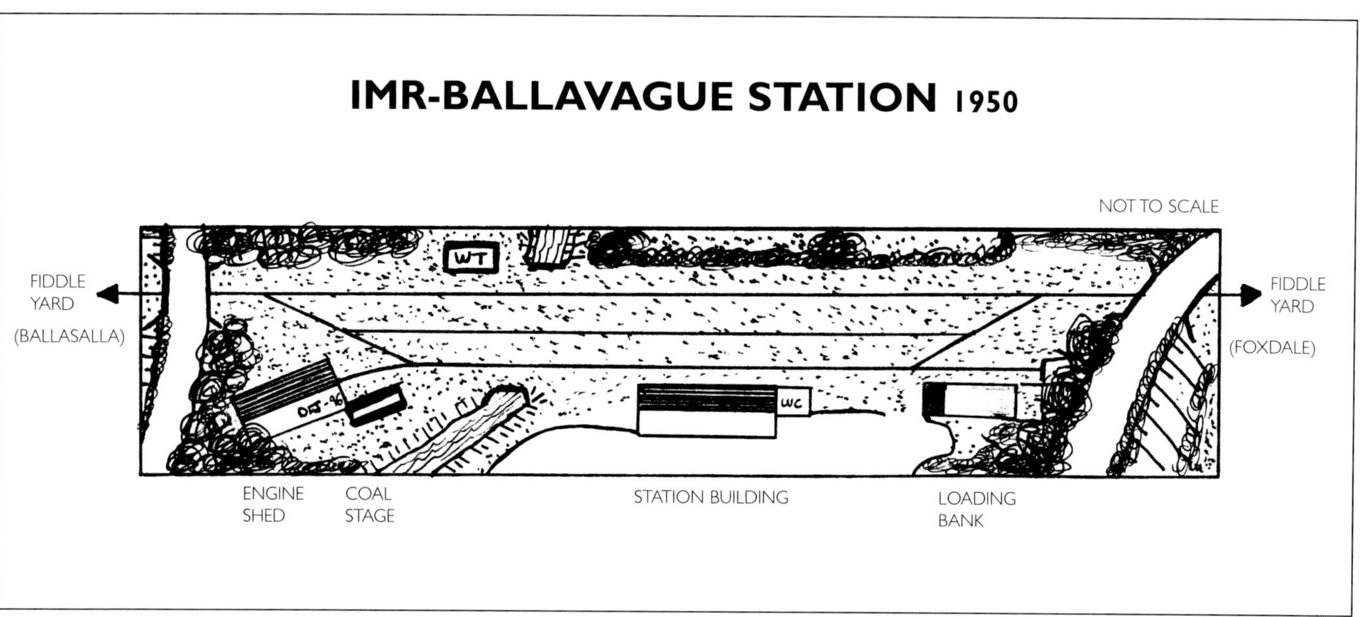

ST GERMAIN'S STATION

St Germain's is basically the Manx Northern Railway's version of the IMR's Santon station. Apart from the MNR's headquarters and terminus at Ramsey, most of the other MNR stations were all of a similar design. All the station buildings, goods sheds and crossing keeper's houses were executed in red sandstone. All stations on the north line were dead straight, again with long passing loops. The exception was St Germain's, which was built on a slight curve that can be exaggerated in model form to make it more interesting.

St Germain's was the last original station before St John's, the IMR and MNR frontier. The opening of Peel Road station four years later in 1883 robbed St Germain's of much of its traffic as Peel Road was much less of a walk for the residents of Peel wishing to use the MNR without having the trouble and extra expense of travelling to St John's on the IMR first. St Germain's soon became a Saturday's only stop on the timetable and the loop and siding lifted.

The loop was restored in 1928 and the siding a year later, when St Germain's came back into fashion as a destination for rail excursions. Once again, both the loop and siding were lifted in the early 1950s. The end of St Germain's station and the whole of the northern line to Ramsey came on September 6 1968.

Although the line has now been closed for nearly 30 years, St Germain's survives virtually intact with just the rails missing. The station building at St Germain's has been a private residence for many years, even before the railway closed.

St Germain's with its very simple track layout and basic facilities could form the scenic part of a simple test track. A basic layout is quicker and cheaper to build and has more chance of being finished before interest begins to flag – as so often happens with much larger layouts. As with all the layout suggestions described in this section, models of the railway buildings will all have to be scratch-built as there are no commercial kits available. However, apart from Douglas station, most of the IMR & MNR buildings are very simple affairs.

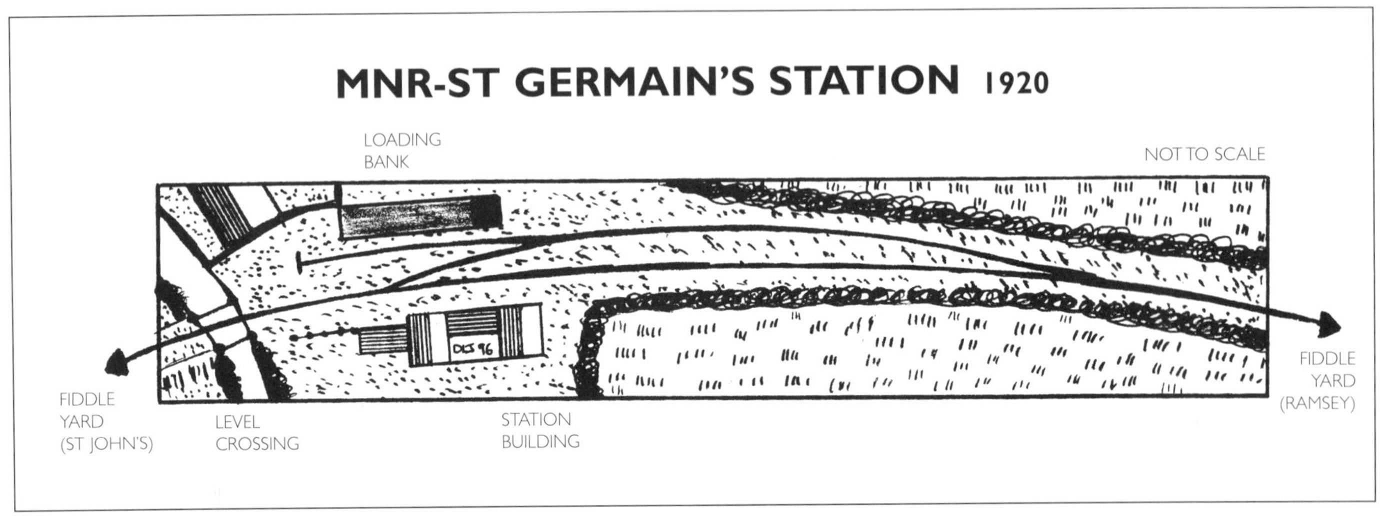

SANTON STATION

This is the second station out of Douglas on the IMR's busy south line. Santon is a delightful little station with a charm all of its own. There are several good reasons as to why this would make an interesting layout. First, the station is completely timeless, the wooden station building dates from the opening of the line in 1874, and apart from the removal of the advertisement hoarding in recent years, not a lot has happened. This allows the modeller to run either period or more modern IMR stock through Santon and still be historically accurate.

The main Douglas to Castletown road bridge forms a natural break at the Port Erin end, while at the Douglas end the line can be curved so it disappears into a group of trees. The advertisement hoarding was used to hide what was known as the 'manure siding', while the grounded body of 4-wheeled luggage/brake van E2 served as a goods shed. There was also a goods platform and a cattle dock.

The close proximity of the main road to Santon station and the introduction of a regular bus service took the local traffic away. In later years, Santon was not included in the timetable. Today it is a request stop only. Once a year, Santon becomes a special destination for hundreds of local children – as Santa lives at Santon! A shuttle service of Santa specials from Douglas brings families to visit Santa's grotto for presents, mince pies and mulled wine.

As with nearly all Isle of Man stations, Santon has palm trees that give it a slightly tropical flavour. The palm trees at Santon are in a row opposite the station building, making it a very attractive setting and a ideal scenic feature for a layout.

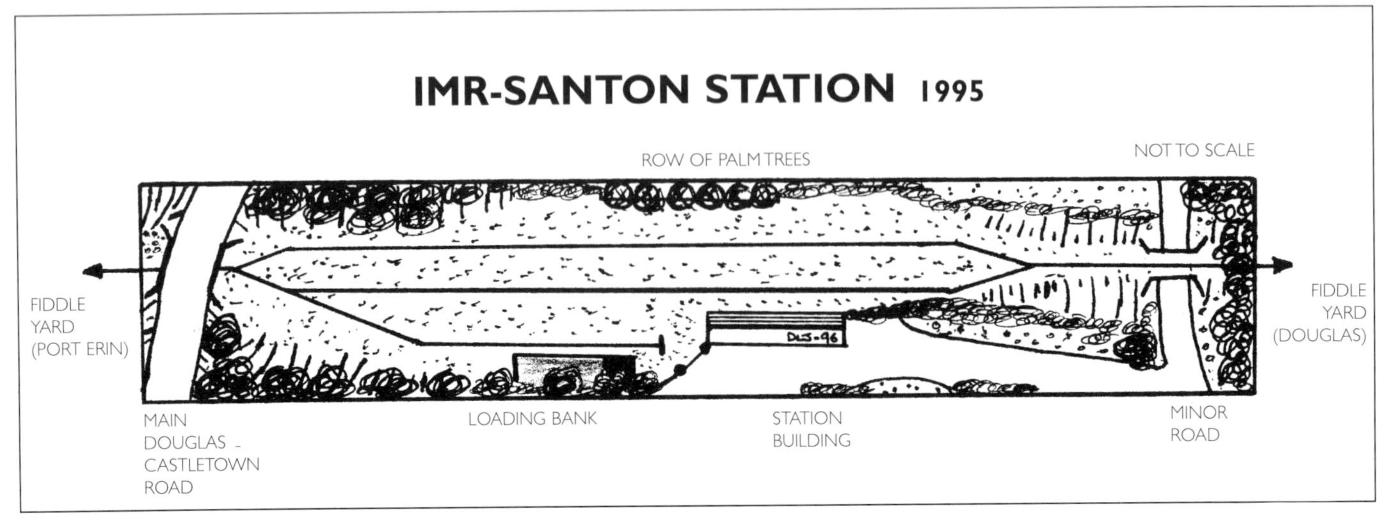

Chapter Eleven

TIMETABLES AND OPERATION

There are two schools of thought on operating a model railway – those who like to run their layout strictly to the prototype timetable and train make-up; while others enjoy just running their trains at will in whatever order that takes their fancy at the time. Down at my local club – The Isle of Man Railway Group – we operate our IMR Castletown layout to prototype practice 95% of the time. The train movements have been taken from a 1950s' IMR timetable to ensure accuracy. However, ever so often, just to keep the fiddle yard operator at the other end of the layout on his toes, an un-prototypical quadruple header with 14 carriages and banker is dispatched. This arouses great interest with the spectators – especially when the train arrives at the other end of the layout, as the fiddle yards are only designed to accommodate six carriages and an engine! Having fun is what railway modelling is all about.

To operate any IMR layout prototypically, a careful study of the various railway timetables is essential. There were normally three types of IMR timetable in any given year. These consisted of the sparse winter timetable followed by an intermediate one at the start of the season normally at Easter. This was superseded by the full summer high season timetable which usually ran from the first week in June to the first or second week in September when it was replaced by the intermediate timetable again until the end of October, when it was back to the winter one once again. On a couple of days each year, the normal IMR service timetable was suspended and a special timetable was introduced for the day. The Manx national holiday on the 5th of July is one such occasion. Tynwald Fair, as it is known, takes place on the village green at St John's. Here the Manx laws are read out both in Manx and English in the grand Tynwald ceremony – Tynwald is the Manx Parliament, which is the oldest government in the world. To get all the people to this event on time, the IMR had to muster every piece of available rolling stock able to carry thousands of passengers. This was a huge logistical exercise for the IMR – all trains had to converge on St John's by 10.00 am. In fact, many of the locomotives and rolling stock such as the south lines No.16 *Mannin*, which for the rest of the year was geographically very conservative, could be found operating on the Peel and Ramsey lines on Tynwald Fair day. A special timetable was also used for the southern and northern agricultural shows held at Castletown and Ramsey. Although, these shows were not on the scale of Tynwald Fair day workings.

In the old days before the 1905 absorption, the Manx Northern Railway meet the Isle of Man Railway at St John's station. The interchange of passengers and goods was a constant source of friction between the two railway companies. It was many years before the IMR allowed the MNR carriages to run on through trains between Ramsey and Douglas.

To take an in-depth look at both the IMR and MNR's timetables and operations would probably fill this book. However, most modellers seem mainly interested in the allocation and operation of the locomotive fleet. As there was no turntables on the Island during the early years of the Isle of Man Railway, all engines out-stationed at Peel, Port Erin and Ramsey, ran chimney first into Douglas – old photographic records show engines Nos. 2, 5 and 8 running in this manner. A 35' long turntable was purchased from Ransome & Rapier in 1924 and situated at St.John's station in the old MNR yard. The table was originally built for the Irish West Clare Railway, who did not require it. It was used by the IMR primarily to turn the coaching stock to allow even weathering on both sides of the carriages to extend the period between repaints. It was at this stage that the locomotives were all turned round to face chimney first out of Douglas. This has remained a tradition of the Manx engines for nearly 50 years. In 1993, during the 'Year of Railways' celebrations, IMR engine No. 4 *Lo ch* was placed back on the rails the other way round following its return to the iMR after operating the 'Steam on Manx Electric Railway' trains. No. 4was to spend the next two years running chimney first into Douglas – defying the old tradition!

Since then a second engine, No. 10 *G. H.Wood*, has been turned around in early April 1996, but only for five days to operate a

steam special for visiting enthusiasts. The IMR plan to turn other engines around in the near future including the ex-MNR 0-6-0T *Caledonia*.

In service, the Manx Peacocks are officially only allowed to haul seven bogie carriages unaided. Trains during the peak summer seasons could be anything up to 13 or 14 coaches long, so double heading and/or banking was and still is normal practice.

Peel and Ramsey trains would very often be combined and double-headed or banked as far as St John's, the main junction station on the Island. At St John's, the combined ex-Douglas trains would be split involving a complicated set of manoeuvres and then go their separate ways to Peel and Ramsey.

During the long winter months, the service was very sparse indeed – trains usually consisted of one or two coaches. In fact, during the early 1960s, when the whole of the IMR was in decline, only one engine in steam would operate the whole system during the winter months! Most engines would be mothballed awaiting the start of the hectic summer season again.

ISLE OF MAN RAILWAY

NOTICE.—The hours or times stated in the Company's Times Tables, Books, Bills and Notices are appointed as those at which it is intended so far as circumstances will permit, that the trains shall depart from and arrive at the several Stations, but their departure or arrival at the times stated is not guaranteed, nor will the Company under any circumstances be held responsible for delay or detention, however occasioned, or any consequences arising therefrom. The right to stop the trains at any station on the lines although not marked as a stopping Station, and to alter or suspend the running of any of the trains, is reserved.

This Time Table will Not Operate on Any Day that may require a Special Time Table.

Time Table for Monday, 3rd July, 1961, and until further notice

DOUGLAS—PORT ERIN LINE — *WEEKDAYS*

To PORT ERIN

Stations		6		8		10		12		14		16 (Runs from 17th July)	
DOUGLAS Dep.	....	10 0	....	10 30	....	11 45	....	2 15	....	3 45	....	5 25	....
Port Soderick ... „	....	....	....	10 41	....	11 56	....	2 26	....	R	....	5 36	....
Ballasalla .... „	....	....	....	10 57	....	12 12	....	2 42	....	4 12	....	5 52	....
Castletown .... „	....	10 33	....	11 3	....	12 18	....	2 48	....	4 18	....	5 58	....
Colby .... „	....	10 42	....	....	....	R	....	R	....	R	....	R	....
Port St. Mary .... „	....	10 49	....	11 18	....	12 33	....	3 3	....	4 33	....	6 13	....
PORT ERIN Arr.	....	10 51	....	11 20	....	12 35	....	3 5	....	4 35	....	6 15	....

To DOUGLAS

Stations	1		7		9		11		13	15 (Runs from 17th July)	17	19 (Runs from 17th July)
PORT ERIN Dep.	7 40	....	10 35	....	11 50	....	1 5	....	2 15	3 50	4 0	5 40
Port St. Mary .... „	7 43	....	10 38	....	11 53	....	1 8	....	2 18	....	4 5	5 43
Colby .... „	7 48	....	10 43	....	11 58	....	1 13	....	2 23	....	4 10	5 48
Castletown .... „	7 57	....	10 52	....	12 7	....	1 22	....	2 32	....	4 19	5 57
Ballasalla .... „	8 3	....	10 58	....	12 13	....	1 28	....	2 41	4 11	4 28	6 3
Port Soderick .... „	8 20	....	R	....	12 30	....	....	....	R	..	4 45	6 20
DOUGLAS Arr.	8 30	....	11 25	....	12 40	....	1 55		3 10	4 40	4 55	6 30

DOUGLAS—PEEL and RAMSEY LINES — *WEEKDAYS*

To PEEL AND RAMSEY

Stations		2 (Not Saturdays)		8	10		12		14		16		
DOUGLAS Dep.	....	....	....	10 25	10 45	....	11 50	....	2 10	....	3 25	....	....
Crosby .... „	....	....	....	R	11 1	....	R	....	R	....	R	....	....
St. John's Arr.	....	....	....	10 48	11 10	....	12 13	....	2 33	....	3 48	....	....
St. John's Dep.	....	....	....	....	11 11	....	12 16	....	2 36	....	3 52	....	....
PEEL .... Arr.	....	....	....	....	11 20	....	12 25	....	2 45	....	4 1	....	....
St. John's Dep.	....	....	....	10 53	....	....	12 18	....	....	....	3 51	....	....
Kirk Michael .... „	....	8 10	....	11 14	....	....	12 39	....	....	....	4 12	....	....
Ballaugh .... „	....	8 18	....	11 22	....	....	12 47	....	....	....	4 20	....	....
Sulby Glen .... „	....	8 24	....	11 28	....	....	12 53	....	—	....	4 26	....	....
Sulby Bridge .... „	....	8 27	....	11 31	....	....	12 56	....	....	....	R	....	....
RAMSEY Arr.	....	8 36	....	11 40		....	1 5	....	....	....	4 37	....	

To DOUGLAS

Stations		1 (Not Saturdays)			7	9		11			17	19 (Runs 5 mins. later to 20th July)	
RAMSEY Dep.	6B45	7 35	....	....	10 0	....	....	1 45	....	....	....	4 0	....
Sulby Bridge .... „	6 55	7 45	....	....	10 10	....	....	1 55	....	....	....	4 10	....
Sulby Glen .... „	6 58	7 48	....	....	10 13	....	....	1 58	....	....	....	4 13	....
Ballaugh .... „	7 4	7 54	....	....	10 19	....	....	2 4	....	....	....	4 19	....
Kirk Michael .... „	7 12	8 1	....	....	10 27	....	....	2 12	....	....	....	4 27	....
St. John's Arr.	CV	....	....		10 47	....	....	2 32	....	....	....	4 47	
PEEL Dep.	7B25	....	8B 5	....	....	12 5	....	2 25	....	....	4 30	....	....
St. John's Arr.	7 34	....	8 14	....	....	12 14	....	2 34	....		4 39	....	....
St. John's Dep.	7B34	....	8B16	....	10 52	....	....	2 37	....	....	4 42	4 52	
Crosby .... „	7 47	....	8 26	....	11 2	....	....	2 47	....	....	4 52	5 2	....
DOUGLAS Arr.	8 0	....	8 40	....	11 15	....	....	3 0	....	....	5 5	5 15	

B — Bus. **R** — Request stop.

CV—Via Cronk-y-Voddy, connecting with bus Ballacraine depart 7-36, due Douglas 8-0.

See also bus time table for connections with steamers.

SUNDAY MORNINGS, commencing 2nd July — TRAINS to KIRK BRADDAN — Douglas depart 10-10 a.m., 10-40 a.m., returning after the Open-air Church Service.

MANXLAND'S MOST CHARMING RESORTS AND LOVELIEST GLENS
are along the routes served by the
STEAM RAILWAY

SPECIAL DAY EXCURSION AND RUNABOUT TICKETS

IT COSTS SO LITTLE TO SEE THE ISLAND BY TRAIN!

DO NOT FAIL TO VISIT—
GLEN WYLLIN *(Kirk Michael Station)*

A beautiful Glen on the Western Shore. BOATING LAKE, CHILDREN'S PLAYGROUND.

Luncheons, Teas, Snacks at reasonable charges.

DOUGLAS. A. M. SHEARD, General Manager.

2542/61—Island Development Co. Ltd.

Typical shed allocation of service engines – summer 1927:

DOUGLAS:	Engines Nos. 2, 5, 8, 9, 13, 15, 16.
RAMSEY:	Engines Nos.1, 7, 14.
PEEL:	Engines Nos. 3, 6.
PORT ERIN:	Engines Nos.10, 12.

Notes:
Engine No. 2 was Douglas station pilot
Engines Nos. 8 & 9 operated on the Peel line
Engines Nos. 5, 13 and 16 operated on the Port Erin line
Engine No. 15 was used for special workings
Engines Nos. 4 & 11 were in Douglas workshops under-going repairs

Typical shed allocation of service engines – summer 1961:

DOUGLAS:	Engines Nos. 5, 11, 12, 13, 14, 15.
RAMSEY:	Engine No. 8.
PEEL:	Nil
PORT ERIN:	Engines No. 10, 16.

Notes:
Engine No.13 was Douglas station pilot.
Engines Nos. 5, 8, 11. operated on the Peel and Ramsey lines
Engines Nos. 10, 12, 16. operated on the Port Erin line
Engines Nos. 10 & 12 spent alterative nights out-stationed at Port Erin.
A couple of engines had been dismantled and others placed into long-term storage at this time. There were a number of standby locomotives, such as engines Nos. 1 & 3 ready to operate specials or stand-in for failed service engines.

Typical shed allocation of service engines – summer 1996:

DOUGLAS:	Engines Nos. 10, 11, 15.
RAMSEY:	None – line closed 1968
PEEL:	None – line closed 1968
PORT ERIN:	Engine No. 12

Notes:
The 1996 timetable (below) only requires three engines with one spare. Withdrawn No. 4 is awaiting a new boiler, as is No.13.

TIMETABLES

MANX ELECTRIC RAILWAY

DAILY FROM MONDAY APRIL 1 – SATURDAY MAY 18 AND MONDAY SEPTEMBER 9 – SUNDAY OCTOBER 27

	AM	NOON	PM	PM	PM		AM	NOON	PM	PM	PM
DOUGLAS Derby Castle	10.00	12.00	2.00	3.00	5.00	RAMSEY	10.00	12.00	1.30	3.30	4.30
LAXEY	10.30	12.30	2.30	3.30	5.30	LAXEY	10.45	12.45	2.15	4.15	5.15
RAMSEY	11.15	1.15	3.15	4.15	6.15	DOUGLAS Derby Castle	11.15	1.15	2.45	4.45	5.45

ENTHUSIASTS WEEK
MAY 18 - 27
CENTENARY FORTNIGHT
JULY 15 - 28
VINTAGE TRANSPORT WEEK
AUGUST 17 - 24
Train and Tram services may be altered during these periods

DAILY FROM SUN MAY 19 – SUN SEPTEMBER 8 | MON – THURS JULY 15 – AUGUST 29 | MON – SAT JULY 15 – AUGUST 31 | WEDS EVENINGS JULY 3 – AUGUST 21 Operated by Illuminated Tram.

	AM	AM	AM	NOON	PM	PM	PM	PM	PM	PM	PM	PM	PM	PM
DOUGLAS Derby Castle	10.00	10.30	11.00	12.00	1.00	1.30	2.00	2.30	3.00	4.00	5.00	5.45	6.00	7.30
GROUDLE													6.10	7.40
LAXEY	10.30	11.00	11.30	12.30	1.30	2.00	2.30	3.00	3.30	4.30	5.30	6.15 STOP	6.30	8.00
RAMSEY	11.15	11.45	12.15	1.15	2.15	2.45	3.15	3.45	4.15	5.15	6.15		7.15	8.45
RAMSEY	—	10.00	11.30	12.00	12.30	1.30	2.30	3.00	3.30	4.00	4.30	5.30	6.30	9.15
LAXEY	9.45	10.45	12.15	12.45	1.15	2.15	3.15	3.45	4.15	4.45	5.15	6.15	7.15	10.00
GROUDLE													7.30	10.15
DOUGLAS Derby Castle	10.15	11.15	12.45	1.15	1.45	2.45	3.45*	4.15	4.45	5.15	5.45	6.45	7.45	10.30

GROUDLE SHUTTLE
Departs **Derby Castle** Quarter to and Quarter past each hour 6.45-8.45pm
Departs **Groudle** on the hour and half hour 7.00-9.00pm

Buses connect with these trams at Derby Castle to or from Douglas Steam Railway Station • ACCORDING TO DEMAND EXTRA TRAMS MAY RUN BETWEEN DOUGLAS AND LAXEY • *Except Mon – Thurs July 15 – Aug 29.
Travel times in mins : **Douglas** – Groudle 12 · Baldrine 18 · Laxey 30 · Dhoon 45 · Ballaglass 55 · Ballajora 63 · Ramsey 75 • **Ramsey** – Ballajora 12 · Ballaglass 20 · Dhoon 30 · Laxey 45 · Baldrine 57 · Groudle 63 · Douglas 75

SNAEFELL MOUNTAIN RAILWAY

ALL SERVICES SUBJECT TO THE WEATHER • JOURNEY TIME 30 MINS

DAILY FROM MONDAY APRIL 29 – SUNDAY SEPTEMBER 29. REGULAR DEPARTURES FROM LAXEY FROM 10.30 AM. LAST GUARANTEED DEPARTURE TO SUMMIT 3.30 PM.

GROUDLE GLEN RAILWAY

OPERATED BY GROUDLE GLEN RAILWAY LIMITED

EASTER – SUN APRIL 7 & MON APRIL 8 • SUNS MAY 5 – SEPT 29 • BANK HOLIDAY MONDAYS MAY 6, MAY 27, AUG 26 • TRAINS OPERATE FROM 11.00 AM TO 4.30 PM
EVENINGS WEDS JULY 3 – AUG 21 TRAINS OPERATE 7.00-9.00 PM. • SPECIAL EVENTS PERIODS – DAILY MAY 23 - 27 & JULY 17 - 24 INCLUSIVE 11.00 AM TO 4.30 PM.

STEAM RAILWAY

BUS CONNECTIONS FOR STEAM RAILWAY FROM MANX ELECTRIC RAILWAY										
DERBY CASTLE	9.29	11.19	1.49	3.49	9.29	10.29	11.19	1.49	3.29	4.29
VILLA MARINA	9.33	11.23	1.53	3.53	9.33	10.33	11.23	1.53	3.33	4.33
RAILWAY STATION	9.45	11.35	2.05	4.05	9.45	10.45	11.35	2.05	3.45	4.45
DOUGLAS TO PORT ERIN	AM	AM	PM	PM	AM	AM	AM	PM	PM	PM
DOUGLAS	10.10	11.45	2.10	4.10	10.10	10.50	11.45	2.10	3.55	4.55
PORT SODERICK	10.23	11.58	2.23	4.23	10.23	11.08	11.58	2.23	4.08	5.08
BALLASALLA	10.45	12.20	2.45	4.45	10.45	11.30	12.20	2.45	4.30	5.30
CASTLETOWN	10.52	12.28	2.52	4.52	10.52	11.37	12.28	2.52	4.37	5.37
COLBY	11.03	12.38	3.03	5.03	11.03	11.48	12.38	3.03	4.48	5.48
PORT ST MARY	11.12	12.47	3.12	5.12	11.12	11.57	12.47	3.12	4.57	5.57
PORT ERIN	11.15	12.50	3.15	5.15	11.15	12.00	12.50	3.15	5.00	6.00
PORT ERIN TO DOUGLAS	AM	PM	PM	PM	AM	PM	PM	PM	PM	PM
PORT ERIN	10.15	12.05	2.15	4.15	10.15	12.05	2.15	3.15	4.15	5.15
PORT ST MARY	10.19	12.09	2.19	4.19	10.19	12.09	2.19	3.19	4.19	5.19
COLBY	10.27	12.17	2.27	4.27	10.27	12.17	2.27	3.27	4.27	5.27
CASTLETOWN	10.38	12.28	2.38	4.38	10.38	12.28	2.38	3.38	4.38	5.38
BALLASALLA	10.45	12.35	2.45	4.45	10.45	12.35	2.45	3.45	4.45	5.45
PORT SODERICK	11.07	12.55	3.07	5.07	11.07	12.55	3.07	4.07	5.07	6.07
DOUGLAS	11.20	1.10	3.20	5.20	11.20	1.10	3.20	4.20	5.20	6.20
BUS CONNECTIONS FOR MANX ELECTRIC RAILWAY FROM STEAM RAILWAY STATION										
RAILWAY STATION	11.25	1.15	3.25	5.25	11.25	1.15	3.25	4.25	5.25	6.25
VILLA MARINA	11.34	1.24	3.34	5.34	11.34	1.24	3.34	4.34	5.34	6.34
DERBY CASTLE	11.38	1.28	3.38	5.38	11.38	1.28	3.38	4.38	5.38	6.38

TRAINS OPERATE ON ALL DATES SHOWN ON THE CALENDAR.
ONLY TRAIN TIMES SHADED GREEN OPERATE ON THE DATES SHADED GREEN IN JULY AND AUGUST.

	M T W T F	SA SU	M T W T F	SA SU	M T W T F	SA SU	M T W T F	SA SU
APRIL	5	6 7	8 9 10 11 12	13 14	15 16 17 18 19	20 21	22 23 24 25 26	27 28
	29 30							
MAY	1 2 3	4 5	6 7 8 9 10	11 12	13 14 15 16 17	18 19	20 21 22 23 24	25 26
	27 28 29 30 31							
JUNE		1 2	3 4 5 6 7	8 9	10 11 12 13 14	15 16	17 18 19 20 21	22 23
	24 25 26 27 28	29 30						
JULY	1 2 3 4 5	6 7	8 9 10 11 12	13 14	15 16 17 18 19	20 21	22 23 24 25 26	27 28
	29 30 31							
AUG	1 2	3 4	5 6 7 8 9	10 11	12 13 14 15 16	17 18	19 20 21 22 23	24 25
	26 27 28 29 30	31						
SEPT		1	2 3 4 5 6	7 8	9 10 11 12 13	14 15	16 17 18 19 20	21 22
	23 24 25 26 27	28 29						
OCT						19 20	21 22 23 24 25	26 27

All trains stop at Santon; Ronaldsway Halt, Level and Ballabeg as required. Passengers wishing to alight must tell the guard on boarding. Passengers wishing to board must give a clear hand signal to the driver.
From **Douglas**: Santon 25 mins; Ronaldsway Halt 39 mins; Ballabeg 47 mins; Level 56 mins. From **Port Erin**: Level 8 mins; Ballabeg 17 mins; Ronaldsway Halt 25 mins; Santon 42 mins.

BIBLIOGRAPHY:

RESEARCH SOURCES:

Obviously, if you are setting out to model any aspect of the Isle of Man Railway, the former Manx Northern railway or any other Manx steam or electric railway for that matter, then almost any book on the subject is well worth reading. Several of the books listed below are currently out of print. The specialist railway booksellers and the second-hand stalls at model railway exhibitions are places to look for old out of print Manx titles. Any of the first four editions of Oakwood's Isle of Man Railway by J. I. C. Boyd are well worth looking for. However, new books are coming along all the time including the new enlarged edition of Mr Boyd's Isle of Man Railway, which Oakwood have published in three volumes, and are considered as essential works of reference for anyone modelling the IMR.

The Isle of Man Railway 1-4 editions – J. I. C. Boyd , Oakwood Press, 1962-1977. ***

The Isle of Man Railway Vol 1-3 – J. I. C. Boyd, Oakwood Press, 1993-1996.

The Isle of Man Railway – Norman Jones, Foxline, 1993.

The Isle of Man Railway – Ian McNab, Green Lake Publications, 1945. ***

The Isle of Man Railway Album – W. J. Wise & J. Joyce, Ian Allan, 1968. ***

The Isle of Man Railway Album – R. Preston Hendry & R. Hendry, David & Charles, 1976. ***

The Railways & Tramways of the Isle of Man – Barry Edwards, OPC, 1993.

The Isle of Man Railways – A Celebration – R. Kirkman & P. Van Zeller, Raven Books, 1993.

The Isle of Man Railway – Gordon Kniveton, Manx Experience, 1987.

The Manx Northern Railway – R. Preston Hendry & R. Hendry, Hillside Publishing, 1980.

On The Isle of Man Narrow Gauge – J. I. C. Boyd, Bradford & Barton, 1978. ***

The Beyer, Peacock Quarterly Review – Vol. 2 April 1928 No. 2.

The Manx Electric Railway Album – R.Preston Hendry & R. Hendry, Hillside Publishing, 1978. ***

The Manx Electric Railway – G. Kniveton, Manx Experience, 1991.

100 Years of the Manx Electric Railway – Keith Pearson, Leading Edge, 1992.

Manx Electric – Mike Goodwyn, Platform 5, 1993.

Goods Rolling Stock of the MER, Douglas Cable Car Group, 1993.

*** = Currently out of print.

Various Isle of Man railway articles in Model Railway News, Model Railway Constructor, Railway Modeller, British Railway Modelling (see separate list), Railway Magazine, Railway World, Steam Railway, Manx Steam Railway News, Manx Transport Review and Modelling Railways Illustrated.

Three very good videos on the Isle of Man Railway are: The Oakwood Video Library No. 3 'Manx Lines Through the Years' and 'The Ivo Peters Collection' Vol. 9 and Vol. 14. All show the IMR in its heyday and are a valuable aid to modelling the Island's railways.

MANX RAILWAY FEATURES IN BRITISH RAILWAY MODELLING

Modelling the Isle of Man Railways	November 1993
'Peveril' re-merges for a face lift	April 1994
Modelling the Isle of Man Railways II	August 1994
IMR 'Peacock' 4mm kit from Branchlines	September 1994
Isle of Man Railway coach kits	November 1994
New Manx Goods	February 1995
The Isle of Man Railway book Volumes 1 & 2	March 1995
The IMR Kipper Vans?	April 1995****
A Manx Electric Railway 4mm scale tram kit	July 1995
The Snaefell Mountain Railway (video)	July 1995
The Groudle Glen Railway	August 1995
Snaefell Mountain Railway (book)	August 1995
More Manx Goods	September 1995
Manx Transport Kaleidoscope (booklet)	September 1995
Manx Modelrail '95 exhibition report	November 1995*
The Upwardly Mobile Snaefell Mountain Railway	December 1995
100 years of the Snaefell Mountain Ry (book)	December 1995
Travelling with the Motorman on the MER video	January 1996
Travelling with the Motorman on the SMR video	January 1996
Live Steam Manx Peacocks	January 1996
More Manx Accessories	February 1996
Snaefell 100 (video)	March 1996 ***
The Ultimate Isle of Man Locomotive Accessory?	April 1996
The Manx 1995 International Railway Festival video	April 1996
Locomotive Profile – The resurrection of 'G.H. Wood'	May 1996 *
Developments at Bradda West	June 1996**

All articles by David Lloyd-Jones
except those marked * = by Cliff Thomas, ** = by John Cox,
*** = by David Brown, **** = by Curtis Devereau
Back issues, subscription details etc. are available from:
British Railway Modelling, The Maltings, West Street, Bourne, Lincolnshire PE10 9PH. Telephone 01778 393313

USEFUL ADDRESSES:

Many IMR locomotives and rolling stock kits are now available from manufacturers in a number of different scales. The names and addresses of the principal specialist suppliers are given below. If you are tempted to have a go at modelling the IMR or even the MNR, it is useful to send for their current catalogues or price lists – but, please enclose a stamped addressed A4 or A5 envelope with your enquiry.

A. P. & M. R. Blackham: 17 Wesley Close, Maidstone, Kent, ME16 9HT Tel: 01622 727379. A growing range of 4mm scale IMR coach lettering, crests and numbering transfers.

Branchlines: P.O. Box 31, Exeter, EX4 6NY. Tel: 01392 437755 Various 4mm scale Manx Peacock 2-4-0Ts and the ex-MNR 0-6-0T *Caledonia* locomotive kits. Also, large 'F' type carriages and a growing range of IMR/MNR freight wagon kits. All can be built to run on either 9 or 12mm gauge track. Also available are scale IMR type chopper couplings and an etched brass fret containing all the IMR and MNR locomotives nameplates, builders plates and chimney numerals.

D. J. B. Engineering: 17 Meadow Way, Bracknell, Berks, RG42 1UE. Tel: 01344 423256. D.J.B. produce a 15mm to 1' scale live steam Manx Peacock 2-4-0T small type, i.e. Nos. 1 to 9 and No. 14, to run on 45mm gauge track. A range of IMR rolling stock is to follow.

Gem Model Railways: 31a Rhos Road, Rhos on Sea, Colwyn Bay, Clwyd, LL28 4RR. The original two 4mm scale white metal IMR locomotives are now back in production. They can be supplied either as just a body kit or with the original cast chassis.

IMP Models: 3 Nightingale Crescent, Birchwood, Lincoln, LN6 0JJ Tel: 01895 231645. Produce a range of semi-scale IMR rolling stock in 16mm scale. Reasonably priced, complete vac-formed kits, which include wheels and couplings.

Malcolm Mills: 5 Bucklow Close, Mooreside, Oldham OL4 2NG. Tel: 01616 529970. Cast resin 4mm scale body kits of the Manx Peacocks Nos. 1 to No. 3 in early condition with small tanks and Salter safety valves, and ex-MNR 0-6-0T *Caledonia*. Both can be supplied to run on either 9 or 12mm gauge track.

The Oakwood Press & Video Library: P.O. Box 122, Headington, Oxford, OX3 8LU. Tel: 01865 874080. Various videos on the Island's railways and, of course, the famous Isle of Man Railway 'Bible' by J. I. C. Boyd in three volumes.

Priory Carriages: 20 Priory Lane, Kents Bank, Grange-over-Sands, Cumbria, LA11 7BH. Produce a range of 15mm to 1' ready-to-run MNR carriages including the 'Foxdale coach' F39 and a MNR luggage /brake. Other IMR/MNR rolling stock to follow soon.

A train leaves Bradda West for Douglas over the crossing leading to the quay on John Cox's OOn3 layout. The locomotive – built from a Gem kit – is No. 8 *Fenella*. Bradda West featured in the June 1966 issue of *BRM*.
JOHN COX

Roxey Mouldings: 58 Dudley Road, Walton on Thames, Surrey, KT12 2JU Tel: 01932 245439. A large range of both IMR and MNR carriage kits in 4mm scale. All kits can be built to run on either 9 or 12mm gauge track. The MNR carriages are also available in 7mm scale on either 16.5 or 21mm gauge track.

There are five narrow gauge modelling societies which support the scale that you are modelling the Isle of Man Railway in:

The 2mm Finescale Association: Membership Secretary, Mr Bill Rakin, 37 Rostle Top Road, Earby, Colne, Lancashire, BB8 6NJ. The Association has within its ranks a narrow gauge group devoted to 2mm to 1' gauge modelling including 3' gauge.

The 009 Society: Membership Secretary, 70 Grove Road, Shirley, Southampton, SO15 3GG. Although 009 is normally associated with 4mm scale models of 2' gauge prototypes running on 9mm gauge. the 009 Society also encompasses other narrow gauges modelled in 4mm scale including 3' gauge.

The 5.5 Scale Association: Hon. Secretary, Malcolm Savage, 28 Whitegates Crescent, Willaston, South Wirral, L64 2UX. The society is devoted to the promotion of 5.5 mm to 1' scale. While the majority of the members work in 12mm gauge, there are a small, but growing number who model 3' prototypes on 16.5mm gauge track.

The 7mm Narrow Gauge Association: Mervyn Axson, Patchway, 12 Hulton Close, Congleton, Cheshire, CW12 3TF. Dedicated to the popular scale of 7mm to 1', which covers any narrow gauge prototype from 15" right up to 3' 6" gauge. This obviously includes 3' gauge running on either 16.5 or 21mm gauge track.

The Association of 16mm Narrow Gauge Modellers: For all those interested in 3' gauge live steam Manx locomotives and rolling stock in the garden running on 45mm gauge track.

Other useful addresses which may help develop your interest in the Isle of Man and its railways are:

Isle of Man Railways: Strathallan Crescent, Douglas, Isle of Man IM2 4NR. Tel: 01624 663366. For up to date news on special events happening on the Island's railway systems – get your name on their mailing list now!

Isle of Man Visitors Club: P.O. Box 25, Dornoch, Sutherland, IV25 3YB. A club for people who appreciate the culture, history and who enjoy visiting the Isle of Man.

The Narrow Gauge Railway Society: Membership Secretary, Peter Salter, 'Wayside' Stibb, Bude, Cornwall EX23 9RG. Provides useful information on all aspects of narrow gauge including 3' gauge systems.

The Isle of Man Steam Railway Supporters Association: Membership Secretary, Peter Hodgitt, 'Lhagagh', The Level, Colby, Isle of Man, IM9 4AG. A 'must' for anybody interested in the steam railways of the Isle of Man. Their quarterly magazine – 'Manx Steam Railway News' covers recent events on both the Isle of Man Railway and the Groudle Glen Railway in detail, combined with historical articles on both systems.

The Manx Electric Railway Society: Membership enquiries to Richard Dodge, 5 Beech Avenue, Onchan, Isle of Man, IM3 3HQ. Another 'must' for anybody interested in Manx railways both past and present. 'The Manx Transport Review' is the Society's quarterly magazine, which not only covers events on all the Island's railways both electric and steam, but also includes other forms of Manx transport, such as buses, coaches, shipping and aircraft – one not to be missed!

HAPPY MANX MODELLING!

Above: A busy scene at Kirk Michael, Barry C. Lane's 5.5mm scale Manx Northern layout, photographed in 1983. An IMR train has arrived from Douglas while a southbound MNR train headed by Sharp, Stewart 2-4-0 *Ramsey* stands in the loop ready to leave with a train of Manx Northern stock down the single line section to St John's. BARRY C. LANE

Front cover:
A train crosses Glen Wyllin viaduct on Barry C. Lane's 5.5mm scale Manx Northern layout. BARRY C. LANE
Inset: The current IMR coat of arms on carriage F15.

Back cover:
A classic IMR scene that oozes atmosphere as Beyer, Peacock No. 12 *Hutchinson* departs Ballasalla station on the remaining south line for Port Erin on September 29 1995. Note the buried track in the station area and the carriage's double footboards to allow passengers to alight at the stations not equipped with platforms.
All photographs by the author except where otherwise acknowledged.